Microelectromechanical Systems: Concepts, Design and Analysis

Microelectromechanical Systems: Concepts, Design and Analysis

Barkley Sparks

Larsen & Keller
www.larsen-keller.com

Microelectromechanical Systems: Concepts, Design and Analysis
Barkley Sparks
ISBN: 978-1-64172-134-9 (Hardback)

Larsen & Keller
Published by Larsen and Keller Education,
5 Penn Plaza,
19th Floor,
New York, NY 10001, USA

Cataloging-in-Publication Data

Microelectromechanical systems : concepts, design and analysis / Barkley Sparks.
p. cm.
Includes bibliographical references and index.
ISBN 978-1-64172-134-9
1. Microelectromechanical systems. 2. Microelectromechanical systems--Design and construction.
3. Electromechanical devices. 4. Microtechnology. I. Sparks, Barkley.
TK7875 .M53 2019
621.381--dc23

Printed and bound in China.

For more information regarding Larsen and Keller Education and its products, please visit the publisher's website www.larsen-keller.com

Table of Contents

Preface

Microelectromechanical systems (MEMS) refer to the technology of microscopic devices that consists of moving parts. It is made up of components, whose size ranges between 1 and 100 micrometers. It ideally consists of a microprocessor and different micro sensors. MEMS can be designed using modified semiconductor device fabrication technologies. Wet etching, dry etching, molding and plating, electro discharge machining are examples of such technologies. Ceramics, metals, silicon and polymers are some of the materials used in MEMS. Some of the basic processes of MEMS are deposition processes, patterning, etching processes and die preparation. This textbook provides comprehensive insights into the study of microelectromechanical systems. It is a compilation of chapters that discuss the most vital concepts with respect to the design and analysis of MEMS. It is an essential guide for both academicians and students who wish to pursue this discipline further.

To facilitate a deeper understanding of the contents of this book a short introduction of every chapter is written below:

Chapter 1, The technology of microscopic devices with moving parts is known as MEMS or microelectromechanical systems. MEMS at nano-scales is termed as nanoelectromechanical systems (NEMS). MEMS combine with micro-optics to form micro-opto-electro-mechanical systems (MOEMS). This chapter introduces briefly the diverse aspects of MEMS, NEMS and MOEMS as well as the basics of smartdust, nanoelectromechanical relay, etc. **Chapter 2**, Microfabrication is the technology of fabricating structures at the micrometer or even smaller scales. The processes of doping, etching, thin films, microlithography, photolithography, etc. are integral to the development of microfabrication technology. The topics included in this chapter on surface micromachining, photolithography, etching, thin-film deposition, wafer dicing, etc. are intended to provide an easy understanding of the varied aspects of microfabrication. **Chapter 3**, Some of the fundamental concepts central to the understanding of microfabrication and MEMS technology are semiconductor conductivity, resistivity, beam theory and beam deflection, etc. The aim of this chapter is to provide a comprehensive understanding of MEMS and includes significant electrical and mechanical concepts pertaining to electrical resistance, conductance, stress and strain, etc. **Chapter 4**, A sensor is a device that is designed to detect and measure changes in its environment and relay the data to other electronics. An actuator is the part of machinery that controls the movement or operation of a system or mechanism. The diverse aspects of sensing and actuation, as well as the working principles of electrostatic-sensitive device, electrostatic sensors, electrostatic actuators, thermal actuators, magnetic actuators, etc. have been discussed in detail in this chapter. **Chapter 5**, Science and technology have undergone rapid developments in the past decade, which has resulted in innovation in significant tools and techniques in the domain of microelectromechanical devices. This chapter explores topics such as microfluidics, bio-MEMS, digital micromirror devices and MEMS gyroscopes, which will provide an insight into MEMS in the modern scenario.

Finally, I would like to thank the entire team involved in the inception of this book for their valuable time and contribution. This book would not have been possible without their efforts. I would also like to thank my friends and family for their constant support.

Barkley Sparks

1

Introduction to Microelectromechanical Systems

The technology of microscopic devices with moving parts is known as MEMS or microelectromechanical systems. MEMS at nano-scales is termed as nanoelectromechanical systems (NEMS). MEMS combine with micro-optics to form micro-opto-electro-mechanical systems (MOEMS). This chapter introduces briefly the diverse aspects of MEMS, NEMS and MOEMS as well as the basics of smartdust, nanoelectromechanical relay, etc.

Microelectromechanical system (MEMS) are mechanical parts and electronic circuits combined to form miniature devices, typically on a semiconductor chip, with dimensions from tens of micrometres to a few hundred micrometres (millionths of a metre). Common applications for MEMS include sensors, actuators, and process-control units.

Interest in creating MEMS grew in the 1980s, but it took nearly two decades to establish the design and manufacturing infrastructure needed for their commercial development. One of the first products with a large market was the automobile air-bag controller, which combines inertia sensors to detect a crash and electronic control circuitry to deploy the air bag in response. Another early application for MEMS was in inkjet printheads. In the late 1990s, following decades of research, a new type of electronic projector was marketed that employed millions of micromirrors, each with its own electronic tilt control, to convert digital signals into images that rival the best traditional television displays. Emerging products include mirror arrays for optical switching in telecommunications, semiconductor chips with integrated mechanical oscillators for radio-frequency applications (such as cellular telephones), and a broad range of biochemical sensors for use in manufacturing, medicine, and security.

MEMS are fabricated by using the processing tools and materials employed in integrated-circuit (IC) manufacturing. Typically, layers of polycrystalline silicon are deposited along with so-called sacrificial layers of silicon dioxide or other materials. The layers are patterned and etched before the sacrificial layers are dissolved to reveal three-dimensional structures, including microscopic cantilevers, chambers, nozzles, wheels, gears, and mirrors. By building these structures with the same batch-processing methods used in IC manufacturing, with many MEMS on a single silicon wafer, significant economies of scale have been achieved. Also, the MEMS components are in essence "built in place," with no subsequent assembly required, in contrast to the manufacture of conventional mechanical devices.

A technical issue in MEMS fabrication concerns the order in which to build the electronic and mechanical omponents. High-temperature annealing is needed to relieve stress and warping of the polycrystalline-silicon layers, but it can damage any electronic circuits that have already been added. On the other hand, building the mechanical components first requires protecting these parts while the electronic circuitry is fabricated. Various solutions have been used, including burying the mechanical parts in shallow trenches prior to the electronics fabrication and then uncovering them afterward.

Barriers to further commercial penetration of MEMS include their cost compared with the cost of simpler technologies, nonstandardization of design and modeling tools, and the need for more reliable packaging. A current research focus is on exploring properties at nanometre dimensions (i.e., at billionths of a metre) for devices known as nanoelectromechanical systems (NEMS). At these scales the frequency of oscillation for structures increases (from megahertz up to gigahertz frequencies), offering new design possibilities (such as for noise filters); however, the devices become increasingly sensitive to any defects arising from their fabrication.

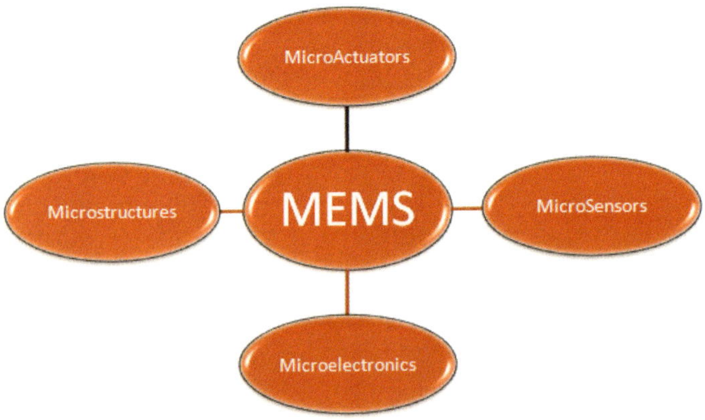

Materials for MEMS Manufacturing

The fabrication of MEMS evolved from the process technology in semiconductor device fabrication, i.e. the basic techniques are deposition of material layers, patterning by photolithography and etching to produce the required shapes.

Silicon

Silicon is the material used to create most integrated circuits used in consumer electronics in the modern industry. The economies of scale, ready availability of inexpensive high-quality materials, and ability to incorporate electronic functionality make silicon attractive for a wide variety of MEMS applications. Silicon also has significant advantages engendered through its material properties. In single crystal form, silicon is an almost perfect Hookean material, meaning that when it is flexed there is virtually no hysteresis and hence almost no energy dissipation. As well as making for highly repeatable motion, this also makes silicon very reliable as it suffers very little fatigue and can have service lifetimes in the range of billions to trillions of cycles without breaking.

Polymers

Even though the electronics industry provides an economy of scale for the silicon industry, crystalline silicon is still a complex and relatively expensive material to produce. Polymers on the other hand can be produced in huge volumes, with a great variety of material characteristics. MEMS devices can be made from polymers by processes such as injection molding, embossing or stereolithography and are especially well suited to microfluidic applications such as disposable blood testing cartridges.

Metals

Metals can also be used to create MEMS elements. While metals do not have some of the advantages displayed by silicon in terms of mechanical properties, when used within their limitations, metals can exhibit very high degrees of reliability. Metals can be deposited by electroplating, evaporation, and sputtering processes. Commonly used metals include gold, nickel, aluminium, copper, chromium, titanium, tungsten, platinum, and silver.

Ceramics

The nitrides of silicon, aluminium and titanium as well as silicon carbide and other ceramics are increasingly applied in MEMS fabrication due to advantageous combinations of material properties. AlN crystallizes in the wurtzite structure and thus shows pyroelectric and piezoelectric properties enabling sensors, for instance, with sensitivity to normal and shear forces. TiN, on the other hand, exhibits a high electrical conductivity and large elastic modulus, making it possible to implement electrostatic MEMS actuation schemes with ultrathin membranes. Moreover, the high resistance of TiN against biocorrosion qualifies the material for applications in biogenic environments and in biosensors.

MEMS Basic Processes

Deposition Processes

One of the basic building blocks in MEMS processing is the ability to deposit thin films of material with a thickness anywhere between one micrometre, to about 100 micrometres. The NEMS process is the same, although the measurement of film deposition ranges from a few nanometres to one micrometre. There are two types of deposition processes, as follows:

Physical Deposition

Physical vapor deposition ("PVD") consists of a process in which a material is removed from a target, and deposited on a surface. Techniques to do this include the process of sputtering, in which an ion beam liberates atoms from a target, allowing them to move through the intervening space and deposit on the desired substrate, and evaporation, in which a material is evaporated from a target using either heat (thermal evaporation) or an electron beam (e-beam evaporation) in a vacuum system.

Chemical Deposition

Chemical deposition techniques include chemical vapor deposition, in which a stream of source gas reacts on the substrate to grow the material desired. This can be further divided into categories depending on the details of the technique, for example, LPCVD (Low Pressure chemical vapor deposition) and PECVD (Plasma-enhanced chemical vapor deposition).

Oxide films can also be grown by the technique of thermal oxidation, in which the (typically silicon) wafer is exposed to oxygen and/or steam, to grow a thin surface layer of silicon dioxide.

Patterning

Patterning in MEMS is the transfer of a pattern into a material.

Lithography

Lithography in MEMS context is typically the transfer of a pattern into a photosensitive material by selective exposure to a radiation source such as light. A photosensitive material is a material that experiences a change in its physical properties when exposed to a radiation source. If a photosensitive material is selectively exposed to radiation (e.g. by masking some of the radiation) the pattern of the radiation on the material is transferred to the material exposed, as the properties of the exposed and unexposed regions differs.

This exposed region can then be removed or treated providing a mask for the underlying substrate. Photolithography is typically used with metal or other thin film deposition, wet and dry etching. Sometimes, photolithography is used to create structure without any kind of post etching. One example is SU8 based lens where SU8 based square blocks are generated. Then the photoresist is melted to form a semi-sphere which acts as a lens.

Electron Beam Lithography

Electron beam lithography (often abbreviated as e-beam lithography) is the practice of scanning a beam of electrons in a patterned fashion across a surface covered with a film (called the resist), ("exposing" the resist) and of selectively removing either exposed or non-exposed regions of the resist ("developing"). The purpose, as with photolithography, is to create very small structures in the resist that can subsequently be transferred to the substrate material, often by etching. It was developed for manufacturing integrated circuits, and is also used for creating nanotechnology architectures.

The primary advantage of electron beam lithography is that it is one of the ways to beat the diffraction limit of light and make features in the nanometer range. This form of maskless lithography has found wide usage in photomask-making used in photolithography, low-volume production of semiconductor components, and research & development.

The key limitation of electron beam lithography is throughput, i.e., the very long time it takes to expose an entire silicon wafer or glass substrate. A long exposure time leaves the user vulnerable to beam drift or instability which may occur during the exposure. Also, the turn-around time for reworking or re-design is lengthened unnecessarily if the pattern is not being changed the second time.

Ion Beam Lithography

It is known that focused-ion beam lithography has the capability of writing extremely fine lines (less than 50 nm line and space has been achieved) without proximity effect. However, because the writing field in ion-beam lithography is quite small, large area patterns must be created by stitching together the small fields.

Ion Track Technology

Ion track technology is a deep cutting tool with a resolution limit around 8 nm applicable to radiation resistant minerals, glasses and polymers. It is capable of generating holes in thin films without any development process. Structural depth can be defined either by ion range or by material thickness. Aspect ratios up to several 10^4 can be reached. The technique can shape and texture materials at a defined inclination angle. Random pattern, single-ion track structures and aimed pattern consisting of individual single tracks can be generated.

X-ray Lithography

X-ray lithography is a process used in electronic industry to selectively remove parts of a thin film. It uses X-rays to transfer a geometric pattern from a mask to a light-sensitive chemical photoresist, or simply "resist", on the substrate. A series of chemical treatments then engraves the produced pattern into the material underneath the photoresist.

Diamond Patterning

A simple way to carve or create patterns on the surface of nanodiamonds without damaging them could lead to a new photonic devices.

Diamond patterning is a method of forming diamond MEMS. It is achieved by the lithographic application of diamond films to a substrate such as silicon. The patterns can be formed by selective deposition through a silicon dioxide mask, or by deposition followed by micromachining or focused ion beam milling.

Etching Processes

There are two basic categories of etching processes: wet etching and dry etching. In the former, the material is dissolved when immersed in a chemical solution. In the latter, the material is sputtered or dissolved using reactive ions or a vapor phase etchant.

Wet Etching

Wet chemical etching consists of selective removal of material by dipping a substrate into a solution that dissolves it. The chemical nature of this etching process provides a good selectivity, which means the etching rate of the target material is considerably higher than the mask material if selected carefully.

Isotropic Etching

Etching progresses at the same speed in all directions. Long and narrow holes in a mask will produce v-shaped grooves in the silicon. The surface of these grooves can be atomically smooth if the etch is carried out correctly, with dimensions and angles being extremely accurate.

Anisotropic Etching

Some single crystal materials, such as silicon, will have different etching rates depending on the crystallographic orientation of the substrate. This is known as anisotropic etching and one of the most common examples is the etching of silicon in KOH (potassium hydroxide), where Si <111> planes etch approximately 100 times slower than other planes (crystallographic orientations). Therefore, etching a rectangular hole in a (100)-Si wafer results in a pyramid shaped etch pit with 54.7° walls, instead of a hole with curved sidewalls as with isotropic etching.

HF Etching

Hydrofluoric acid is commonly used as an aqueous etchant for silicon dioxide (SiO_2, also known as BOX for SOI), usually in 49% concentrated form, 5:1, 10:1 or 20:1 BOE (buffered oxide etchant) or BHF (Buffered HF). They were first used in medieval times for glass etching. It was used in IC fabrication for patterning the gate oxide until the process step was replaced by RIE.

Hydrofluoric acid is considered one of the more dangerous acids in the cleanroom. It penetrates the skin upon contact and it diffuses straight to the bone. Therefore, the damage is not felt until it is too late.

Electrochemical Etching

Electrochemical etching (ECE) for dopant-selective removal of silicon is a common method to automate and to selectively control etching. An active p-n diode junction is required, and either type of dopant can be the etch-resistant ("etch-stop") material. Boron is the most common etch-stop dopant. In combination with wet anisotropic etching as described above, ECE has been used successfully for controlling silicon diaphragm thickness in commercial piezoresistive silicon pressure sensors. Selectively doped regions can be created either by implantation, diffusion, or epitaxial deposition of silicon.

Vapor Etching

Xenon Difluoride

Xenon difluoride (XeF_2) is a dry vapor phase isotropic etch for silicon originally applied for MEMS in 1995 at University of California, Los Angeles. Primarily used for releasing metal and dielectric structures by undercutting silicon, XeF_2 has the advantage of a stiction-free release unlike wet etchants. Its etch selectivity to silicon is very high, allowing it to work with photoresist, SiO_2, silicon nitride, and various metals for masking. Its reaction to silicon is "plasmaless", is purely chemical and spontaneous and is often operated in pulsed mode. Models of the etching action are available, and university laboratories and various commercial tools offer solutions using this approach.

Plasma Etching

Modern VLSI processes avoid wet etching, and use plasma etching instead. Plasma etchers can operate in several modes by adjusting the parameters of the plasma. Ordinary plasma etching operates between 0.1 and 5 Torr. (This unit of pressure, commonly used in vacuum engineering, equals approximately 133.3 pascals.) The plasma produces energetic free radicals, neutrally charged, that react at the surface of the wafer. Since neutral particles attack the wafer from all angles, this process is isotropic.

Plasma etching can be isotropic, i.e., exhibiting a lateral undercut rate on a patterned surface approximately the same as its downward etch rate, or can be anisotropic, i.e., exhibiting a smaller lateral undercut rate than its downward etch rate. Such anisotropy is maximized in deep reactive ion etching. The use of the term anisotropy for plasma etching should not be conflated with the use of the same term when referring to orientation-dependent etching.

The source gas for the plasma usually contains small molecules rich in chlorine or fluorine. For instance, carbon tetrachloride (CCl4) etches silicon and aluminium, and trifluoromethane etches silicon dioxide and silicon nitride. A plasma containing oxygen is used to oxidize ("ash") photoresist and facilitate its removal.

Ion milling, or sputter etching, uses lower pressures, often as low as 10–4 Torr (10 mPa). It bombards the wafer with energetic ions of noble gases, often Ar+, which knock atoms from the substrate by transferring momentum. Because the etching is performed by ions, which approach the wafer approximately from one direction, this process is highly anisotropic. On the other hand, it tends to display poor selectivity. Reactive-ion etching (RIE) operates under conditions intermediate between sputter and plasma etching (between 10–3 and 10–1 Torr). Deep reactive-ion etching (DRIE) modifies the RIE technique to produce deep, narrow features.

Sputtering

Reactive Ion Etching (RIE)

In reactive-ion etching (RIE), the substrate is placed inside a reactor, and several gases are introduced. A plasma is struck in the gas mixture using an RF power source, which breaks the gas molecules into ions. The ions accelerate towards, and react with, the surface of the material being etched, forming another gaseous material. This is known as the chemical part of reactive ion etching. There is also a physical part, which is similar to the sputtering deposition process. If the ions have high enough energy, they can knock atoms out of the material to be etched without a chemical reaction. It is a very complex task to develop dry etch processes that balance chemical and physical etching, since there are many parameters to adjust. By changing the balance it is possible to influence the anisotropy of the etching, since the chemical part is isotropic and the physical part highly anisotropic the combination can form sidewalls that have shapes from rounded to vertical.

Deep RIE (DRIE) is a special subclass of RIE that is growing in popularity. In this process, etch depths of hundreds of micrometres are achieved with almost vertical sidewalls. The primary technology is based on the so-called "Bosch process", named after the German company Robert Bosch, which filed the original patent, where two different gas compositions alternate in the reactor. Currently there are two variations of the DRIE. The first variation consists of three distinct steps (the original Bosch process) while the second variation only consists of two steps.

In the first variation, the etch cycle is as follows:

(i) SF_6 isotropic etch;

(ii) C_4F_8 passivation;

(iii) SF_6 anisoptropic etch for floor cleaning.

In the 2nd variation, steps (i) and (iii) are combined.

Both variations operate similarly. The C_4F_8 creates a polymer on the surface of the substrate, and the second gas composition (SF_6 and O_2) etches the substrate. The polymer is immediately sputtered away by the physical part of the etching, but only on the horizontal surfaces and not the sidewalls. Since the polymer only dissolves very slowly in the chemical part of the etching, it builds up on the sidewalls and protects them from etching. As a result, etching aspect ratios of 50 to 1 can be achieved. The process can easily be used to etch completely through a silicon substrate, and etch rates are 3–6 times higher than wet etching.

Die Preparation

After preparing a large number of MEMS devices on a silicon wafer, individual dies have to be separated, which is called die preparation in semiconductor technology. For some applications, the separation is preceded by wafer backgrinding in order to reduce the wafer thickness. Wafer dicing may then be performed either by sawing using a cooling liquid or a dry laser process called stealth dicing.

MEMS Manufacturing Technologies

Bulk Micromachining

Bulk micromachining is the oldest paradigm of silicon based MEMS. The whole thickness of a silicon wafer is used for building the micro-mechanical structures. Silicon is machined using various etching processes. Anodic bonding of glass plates or additional silicon wafers is used for adding features in the third dimension and for hermetic encapsulation. Bulk micromachining has been essential in enabling high performance pressure sensors and accelerometers that changed the sensor industry in the 1980s and 90's.

Surface Micromachining

Surface micromachining uses layers deposited on the surface of a substrate as the structural materials, rather than using the substrate itself. Surface micromachining was created in the late 1980s to render micromachining of silicon more compatible with planar integrated circuit technology, with the goal of combining MEMS and integrated circuits on the same silicon wafer. The original surface micromachining concept was based on thin polycrystalline silicon layers patterned as movable mechanical structures and released by sacrificial etching of the underlying oxide layer. Interdigital comb electrodes were used to produce in-plane forces and to detect in-plane movement capacitively. This MEMS paradigm has enabled the manufacturing of low cost accelerometers for e.g. automotive air-bag systems and other applications where low performance and/or high g-ranges are sufficient. Analog Devices has pioneered the industrialization of surface micromachining and has realized the co-integration of MEMS and integrated circuits.

High Aspect Ratio (HAR) Silicon Micromachining

Both bulk and surface silicon micromachining are used in the industrial production of sensors, ink-jet nozzles, and other devices. But in many cases the distinction between these two has diminished. A new etching technology, deep reactive-ion etching, has made it possible to combine good performance typical of bulk micromachining with comb structures and in-plane operation typical of surface micromachining. While it is common in surface micromachining to have structural layer thickness in the range of 2 μm, in HAR silicon micromachining the thickness can be from 10 to 100 μm. The materials commonly used in HAR silicon micromachining are thick polycrystalline silicon, known as epi-poly, and bonded silicon-on-insulator (SOI) wafers although processes for bulk silicon wafer also have been created (SCREAM). Bonding a second wafer by glass frit bonding, anodic bonding or alloy bonding is used to protect the MEMS structures. Integrated circuits are typically not combined with HAR silicon micromachining.

Microelectromechanical systems chip, sometimes called "lab on a chip"

Applications

A Texas Instruments DMD chip for cinema projection

Some common commercial applications of MEMS include:

- Inkjet printers, which use piezoelectrics or thermal bubble ejection to deposit ink on paper.
- Accelerometers in modern cars for a large number of purposes including airbag deployment and electronic stability control.

- Inertial measurement units (IMUs): MEMS Accelerometers and MEMS gyroscopes in remote controlled, or autonomous, helicopters, planes and multirotors (also known as drones), used for automatically sensing and balancing flying characteristics of roll, pitch and yaw. MEMS magnetic field sensor (magnetometer) may also be incorporated in such devices to provide directional heading. MEMS are also used in Inertial navigation systems (INSs) of modern cars, airplanes, submarines and other vehicles to detect yaw, pitch, and roll; for example, the autopilot of an airplane.
- Accelerometers in consumer electronics devices such as game controllers (Nintendo Wii), personal media players / cell phones (virtually all smartphones, various HTC PDA models) and a number of Digital Cameras (various Canon Digital IXUS models). Also used in PCs to park the hard disk head when free-fall is detected, to prevent damage and data loss.
- MEMS microphones in portable devices, e.g., mobile phones, head sets and laptops. The market for smart microphones includes smartphones, wearable devices, smart home and automotive applications.
- Silicon pressure sensors e.g., car tire pressure sensors, and disposable blood pressure sensors
- Displays e.g., the digital micromirror device (DMD) chip in a projector based on DLP technology, which has a surface with several hundred thousand micromirrors or single micro-scanning-mirrors also called microscanners
- Optical switching technology, which is used for switching technology and alignment for data communications
- Bio-MEMS applications in medical and health related technologies from Lab-On-Chip to MicroTotalAnalysis (biosensor, chemosensor), or embedded in medical devices e.g. stents.
- Interferometric modulator display (IMOD) applications in consumer electronics (primarily displays for mobile devices), used to create interferometric modulation – reflective display technology as found in mirasol displays
- Fluid acceleration such as for micro-cooling
- Micro-scale energy harvesting including piezoelectric, electrostatic and electromagnetic micro harvesters.
- Micromachined ultrasound transducers.

Industry Structure

The global market for micro-electromechanical systems, which includes products such as automobile airbag systems, display systems and inkjet cartridges totaled $40 billion in 2006 according to Global MEMS/Microsystems Markets and Opportunities, a research report from SEMI and Yole Developpement and is forecasted to reach $72 billion by 2011.

Companies with strong MEMS programs come in many sizes. Larger firms specialize in manufacturing high volume inexpensive components or packaged solutions for end markets such as automobiles, biomedical, and electronics. Smaller firms provide value in innovative solutions and absorb the expense of custom fabrication with high sales margins. Both large and small companies typically invest in R&D to explore new MEMS technology.

The market for materials and equipment used to manufacture MEMS devices topped $1 billion worldwide in 2006. Materials demand is driven by substrates, making up over 70 percent of the market, packaging coatings and increasing use of chemical mechanical planarization (CMP). While MEMS manufacturing continues to be dominated by used semiconductor equipment, there is a migration to 200 mm lines and select new tools, including etch and bonding for certain MEMS applications.

Advantages and Challenges of MEMS

The small size of MEMS is attractive for many applications because feature sizes are typically as small as 1 micrometer or less. Hence, for optical applications, features may be made with size on the order of the wavelength of light. Their small size also allows applications to be developed which would otherwise be impossible. For example, micromechanical switches fabricated as part of a communications circuit allow phase shifting and signal switching at speeds that would be impossible to achieve using macro-scale switches. To illustrate the scale of a typical microsystem, Figure shows a micromachined mirror assembly next to a spider mite. The mirror assembly is about 100 micrometer wide, and it is dwarfed by the spider mite. The mechanical devices surrounding the mirror allow it to be positioned accurately as part of an optical network.

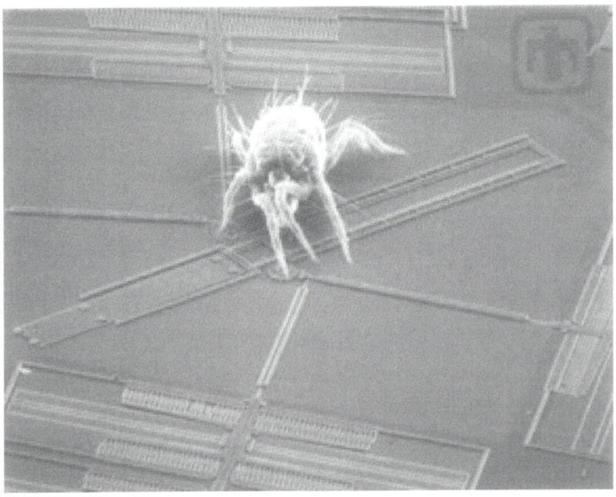

A spider mite dwarfs a micromachined mirror assembly

Other advantages include the on-chip integration of electromechanical systems and the circuitry used to control them, allowing further miniaturization. Furthermore, many MEMS fabrication technologies allow parallel fabrication of thousands of systems by leveraging the parallel fabrication techniques of the integrated circuit industry. This may lead to a reduction in the manufacturing cost and improvement in reliability.

Like any technology, microsystems present some challenges. Because micromechanisms operate at a size scale far below that of typical mechanical devices, surface forces such as adhesion and friction may dominate over other forces in the system, leading to failure of the device. Careful device design, fabrication, and testing are required to reduce these effects. Their small size also makes it difficult to interact mechanically with MEMS components. In many microelectromechanical systems, electrical or optical signals are used to interface with, provide power to, and control the device, rather than manual, hydraulic, or pneumatic control typically seen in macro-scale mechanical devices. In addition, packaging of MEMS components has often presented a challenge because each device must be packaged in a way that keeps the components clean and free from contamination, while also allowing mechanical motion and, in many cases, interaction with the environment. For example, a MEMS pressure sensor requires a package that exposes the sensor to the ambient pressure while protecting the electronic circuitry from dust or other particles. Finally, while parallel fabrication techniques can reduce the manufacturing cost of many units, MEMS development may be more costly because few units are produced at a time using complex and expensive fabrication equipment.

Nanoelectromechanical Systems

Nanoelectromechanical systems (NEMS) are made of electromechanical devices that have critical dimensions from hundreds to a few nanometers. By exploring nanoscale effects, NEMS present interesting and unique characteristics, which deviate greatly from their predecessor microelectromechanical systems (MEMS). For instance, NEMS-based devices can have fundamental frequencies in microwave range (~100 GHz) ; mechanical quality factors in the tens of thousands, meaning low-energy dissipation; active mass in the femtogram range; force sensitivity at the attonewton level; mass sensitivity up to attogram and subattogram levels; heat capacities far below a "yoctocalorie" ; power consumption in the order of 10 attowatts ; extreme high integration level, approaching 1012 elements per square centimeter. All these distinguished properties of NEMS devices pave the way to applications such as force sensors, chemical sensors, biological sensors, and ultra high frequency resonators.

The interesting properties of the NEMS devices typically arise from the behavior of the active parts, which, in most cases, are in the forms of cantilevers or doubly clamped beams with dimensions at nanometer scale. The materials for those active components include silicon and silicon carbide, carbon nanotubes, and gold and platinum, to name a few. Silicon is the basic material for integrated circuit (IC) technology during the past few decades, and MEMS and is widely used to build NEMS. Figure is a scanning electron microscopy (SEM) image of a double-clamped resonator fabricated from a bulk, single-crystal silicon substrate. However, ultrasmall silicon-based NEMS fail to achieve desired high-quality factors because of the dominance of surface effects, such as surface oxidation and reconstruction, and thermoelastic damping. Limitations in strength and flexibility also compromise the performance of silicon-based NEMS actuators. Instead, carbon nanotubes can well represent the ideas of NEMS, given their nearly one-dimensional structures with high-aspect ratio, perfect terminated surfaces, and excellent electrical and mechanical properties. Because of significant advances in growth, manipulation, and knowledge of electrical and mechanical properties, carbon nanotubes have become the most promising building blocks for the next generation of NEMS. Several carbon nanotube–based functional NEMS devices have been reported so far. Similar to carbon nanotubes, nanowires are another type of onedimensional novel nanostructure for building NEMS because of their size and controllable electrical properties.

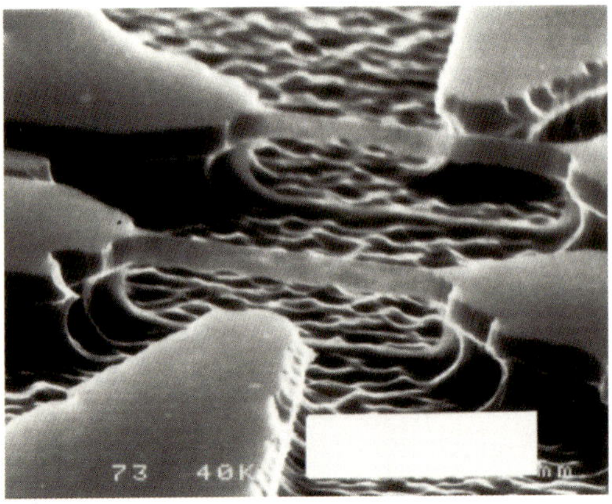

SEM Image of an undercut Si beam, with length of 7.7 m, width of 0.33 m and height of 0.8 m

Nanoelectromechanical Systems

Carbon Nanotubes

Carbon nanotubes exist as a macromolecule of carbon, analogous to a sheet of graphite rolled into a cylinder. They were discovered by Sumio Iijima in 1991 and are a subset of the family of fullerene structures. The properties of the nanotubes depend on the atomic arrangement (how sheets of graphite are rolled to form a cylinder), their diameter, and their length. They are light, stiff, flexible, thermally stable, and chemically inert. They have the ability to be either metallic or semiconducting depending on the "twist" of the tube, which is called "chirality" or "helicity." Nanotubes may exist as either single-walled or multiwalled structures. Multiwalled carbon nanotubes (MWNTs) are simply composed of multiple concentric single-walled carbon nanotubes. The spacing between the neighboring graphite layers in MWNTs is ~0.34 nm. These layers interact with each other via van der Waals forces.

The methods to synthesize carbon nanotubes include electric arc-discharge, laser ablation, and catalytic chemical vapor deposition (CVD) methods. During synthesis, nanotubes are usually mixed with residues, including various types of carbon particles.

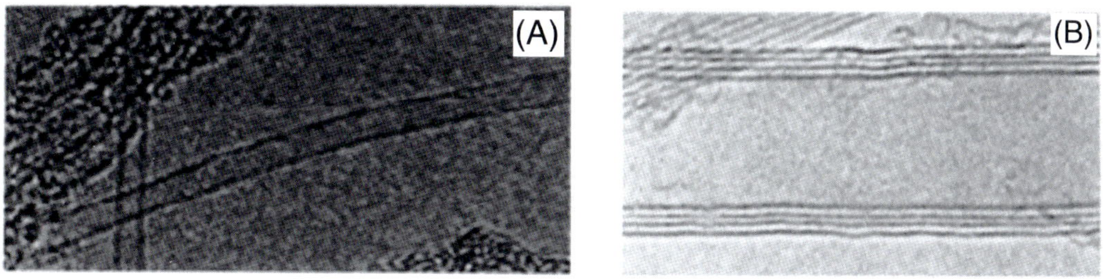

High-resolution transmission electron microscopy image of typical single-walled carbon nanotubes (SWNT) (A) and multi-walled carbon nanotubes (MWNT) (B).

For most applications and tests, a purification process is required. In one of the most common approaches, nanotubes are ultrasonically dispersed in a liquid (e.g., isopropanol) and the suspension is centrifuged to remove large particles. Other methods, including dielectrophoretic separation, are being developed to provide improved yield.

The mechanical and electrical properties of carbon nanotubes have been under intensive study during the past decade. The electronics of carbon nanotubes is extensively reviewed by McEuen. Besides, the study of the coupled electromechanical properties, which are essential to NEMS, is rapidly progressing. Some interesting results have been reported regarding the fact that the electrical properties of carbon nanotubes are sensitive to the structure variation and can be changed dramatically because of the change of the atomic bonds induced by mechanical deformations. It is known that carbon nanotubes can even change from metallic to semiconducting when subjected to mechanical deformation.

Fabrication Methods

The fabrication processes of NEMS devices can be categorized according to two approaches. Top-down approaches, that evolved from manufacturing of MEMS structures, use submicron lithographic techniques, such as electron-beam lithography, to fabricate structures from bulk materials, either thin films or bulk substrates. Bottom-up approaches fabricate the nanoscale devices by sequentially assembling of atoms and molecules as building blocks. Top-down fabrication is size limited by facts such as the resolution of the electron-beam lithography, etching-induced roughness, and synthesis constraints in epitaxially grown substrates. Significant interest has been shown in the integration of nanoscale materials such as carbon nanotubes and nanowires, fabricated by bottom-up approaches, to build nanodevices. Most of the nanodevices reported so far in the literature are obtained by "hybrid" approaches, that is, combination of bottom-up (self assembly) and top-down (lithographic) approaches.

One of the key and most challenging issues of building carbon nanotubes–based or nanowires-based NEMS is the positioning of nanotubes or nanowires at the desired locations with high accuracy and high throughput. Reported methods of manipulation and positioning of nanotubes are briefly summarized in the following topic.

Random Dispersion Followed by e-beam Lithography

After purification, a small aliquot of a nanotube suspension is deposited onto a substrate. The result is nanotubes randomly dispersed on the substrate. Nanotubes on the substrate are imaged inside a scanning electron microscope (SEM) and then this image is digitized and imported to a mask-drawing software, where the mask for the subsequent electronbeam lithography is designed. In the mask layout, pads are designed to superimpose over the carbon nanotubes. Wet etching is employed to remove the material under the carbon nanotubes to form freestanding nanotube structures. This process requires an alignment capability in the lithographic step with an accuracy of 0.1 m or better. This method was firstly employed to make nanotube structures for mechanical testing. The reported NEMS devices based on this method include nanotube–based rotational actuators and nanowire-based resonators.

Nanomanipulation

Manipulation of individual carbon nanotubes using piezo-driven manipulators inside electron microscope chambers is one of the most commonly used methods to build NEMS and structures for mechanical testing. In general, the manipulation and positioning of nanotubes is accomplished in the following manner: a source of nanotubes is positioned close to the manipulator inside the microscope; the manipulator probe is moved close to the nanotubes under visual surveillance of the microscope monitor until a protruding nanotube is attracted to the manipulator due to either van der Waals forces or electrostatic forces; the free end of the attracted nanotube is "spot welded" by the electron-beam-induced deposition (EBID) of hydrocarbon or metals, like platinum from adequate precursor gases.

Shows a three-dimensional nanomanipulator (Klocke Nanotechnik Co.) having the capability of moving in X, Y, and Z directions with nanometer displacement resolution. The manipulation process of an individual carbon nanotube is illustrated.

External Field Alignment

DC/AC electric fields have been successfully used in the manipulation of nanowires, nanotubes, and bioparticles. Microfabricated electrodes are typically used to create an electric field in the gap between them. A droplet containing carbon nanotubes in suspension is dispensed into the gap with a micropipette. The applied electric field aligns the nanotubes, due to the dielectrophoretic effect, which results in the bridging of the electrodes by a single nanotube. The voltage drop that arises when the circuit is closed (DC component) ensures the manipulation of only one nanotube. Besides, AC dielectrophoresis has been employed to successfully separate metallic from semiconducting single-walled carbon nanotubes in suspension. NEMS devices fabricated using this method include nanotube–based nanorelays.

Huang et al. demonstrated another method for aligning nanowires. A laminar flow was employed to achieve preferential orientation of nanowires on chemically patterned surfaces. This method was successfully used in the alignment of silicon nanowires. Magnetic fields have also been used to align carbon nanotubes.

Direct Growth

Instead of manipulating and aligning carbon nanotubes after their manufacturing, researchers have also examined methods for controlled direct growth. Huang et al. used the microcontact printing technique to directly grow aligned nanotubes vertically.

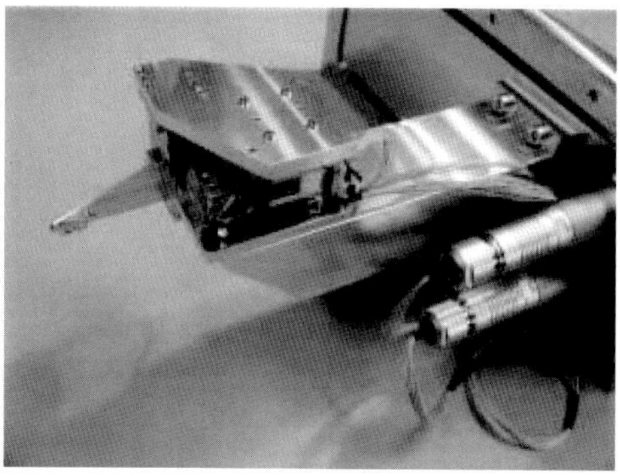

Klocke Nanotechnik nanomanipulator possessing nanometer resolution in the x, y, and z axes.

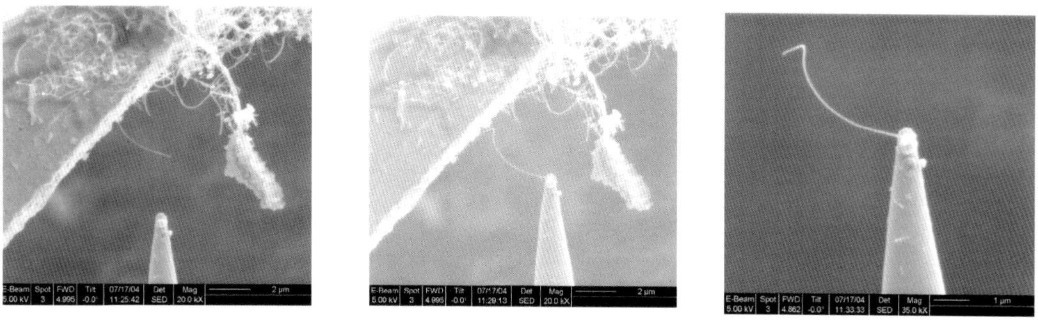

(A) (B) (C)

SEM images of the manipulation of carbon nanotubes using the three-dimensional Klocke Nanotechnik nanomanipulator.

(A) Manipulator probe is approaching a protruding nanotube. The sample is dried nanotube solution on top of a TEM copper grid. (B) Manipulator probe makes contact with the free end of the nanotube and the nanotube is welded to the probe by EBID of platinum. (C) A single nanotube mounted to the manipulator probe.

Dai et al. reported several patterned growth approaches developed in his group. The idea is to pattern the catalyst in an arrayed fashion and control the growth of carbon nanotubes from specific catalytic sites. The authors successfully carried out patterned growth of both MWNTs and SWNTs and exploited methods including self-assembly and external electricfield control. Figure below shows a SEM image of suspended single-walled nanotubes grown by electric-field-directed CVD method. The carbon nanotube–based tunable oscillators, reported in Ref. were fabricated using this method.

Self-Assembly

Self-assembly is a method of constructing nanostructures by forming stable bonds between the organic or nonorganic molecules and substrates. Recently, Rao et al. reported an approach in large-scale assembly of carbon nanotubes with high throughput. Dip Pen Nanolithography (DPN), a technique invented by Mirkin's group, was employed to functionalize the specific surface regions either with polar chemical groups such as amino ($-NH_2/-NH^+{}_3$) or carboxyl ($-COOH/-COO^-$), or with nonpolar groups such as methyl ($-CH_3$. When the substrate with functionalized surfaces was introduced into a liquid suspension of carbon nanotubes, the nanotubes were attracted toward the polar regions and self-assembled to form predesigned structures, usually within 10 s, with a yield higher than 90%. The reported method is scalable to large arrays of nanotube devices by using high-throughput patterning methods such as photolithography, stamping, or massively parallel DPN.

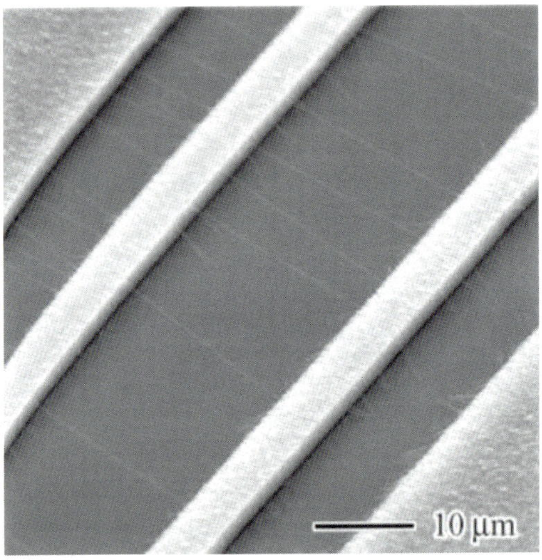

Electric-field-directed freestanding single-walled nanotubes

Inducing and Detecting Motion

For nanostructures, both inducing and detecting motion are challenging. Some of the methods routinely used in MEMS face challenges when the size shrinks from microscale to nanoscale. For example, optical methods, such as simple-beam deflection schemes or more sophisticated optical and fiber-optical interferometry—both commonly used in scanning probe microscopy to detect the deflection of the cantilevers—generally fall beyond the diffraction limit, which means these methods cannot be applied to objects with cross-section much smaller than the wavelength of light.

Inducting Motion

Similar to MEMS, electrostatic actuation of nanostructures by an applied electrical field is commonly used for the actuation of NEMS (e.g., nanotweezers). The Lorenz force has been used to move small conducting beams, with alternating currents passing through them in a strong transverse magnetic field. The induced electromotive force, or voltage, can be detected as a measure of the motion. This method requires a fully conducting path and works well with a beam clamped at both ends. Other actuation methods include piezoelectric actuation, thermal actuation using bilayers of materials with different thermal expansion, thermal in-plane actuation due to a specially designed topography, and scanning tunneling microscope.

Detecting Motion

The most straightforward method is by direct observation of the motion under electron microscopes. This visualization method, typically with resolution in the nanometer scale, projects the motion in the direction to be perpendicular to the electron beam. Limitations in depth of focus requires that the nano-object motion be primarily in a plane, which normally is coaxial with the electron beam. Electron tunneling is a very sensitive method that can detect subnanometer motion by the exponential dependence of the electron tunneling current on the separation between tunneling electrodes. Therefore, this technique is widely used in NEMS motion detection. Magnetomotive detection is a method based on the presence of an electrostatic field, either uniform or spatially inhomogeneous, through which a conductor is moved. The time-varying flux generates an induced electromotive force in the loop, which is proportional to the motion. The displacement detection sensitivity of this technique is less than 1 Å. It is known that carbon nanotubes can act as transistors; as such they can be used to sense their own motion. Capacitance sensors have been widely used in MEMS. They can also be used in NEMS motion sensing with a resolution of a few nanometers, and the resolution can be potentially increased to Angstrom range provided that the capacitance measurement can be improved by one order of magnitude.

Functional Nanoelectromechanical Systems Devices

Carbon Nanotube-based Nanoelectromechanical Systems Devices

Nonvolatile Random Access Memory- A carbon nanotube–based nonvolatile random access memory (NRAM) reported by Reukes et al. is illustrated in Fig. The device is a suspended SWNT crossbar array for both I/O and switchable, bistable device elements with well-defined OFF and ON states. This crossbar consists of a set of parallel SWNTs or nanowires (lower) on a substrate composed of a conducting layer (e.g., highly doped silicon [dark gray]) that terminates in a thin dielectric layer (e.g., SiO2 [light gray]) and a set of perpendicular SWNTs (upper) that are suspended on a periodic array of inorganic or organic supports. Each nanotube is contacted by a metal electrode. Each cross-point in this structure corresponds to a device element with a SWNT suspended above a perpendicular nanoscale wire. Qualitatively, bistability can be envisioned as arising from the interplay of the elastic energy and the van der Waals energy when the upper nanotube is freestanding or the suspended SWNT is deflected and in contact with the lower nanotube. Because the nanotube junction resistance depends exponentially on the separation gap, the separated upper-to-lower nanotube junction resistance will be orders of magnitude higher than that of the contact junction. Therefore, two states—OFF and ON—are well defined. For a device element, these two states can be read easily by measuring the resistance of the junction and, moreover, can be switched between OFF and ON states by applying voltage pulses to nanotubes at corresponding electrodes to produce attractive or repulsive electrostatic forces. A key aspect of this device is that the separation between top and bottom conductors must in the order of 10 nm. In such case, the van der Waals energy overcomes the elastic energy when the junction is actuated (ON state) and remains on this state even if the electrical field is turned off (nonvolatile feature).

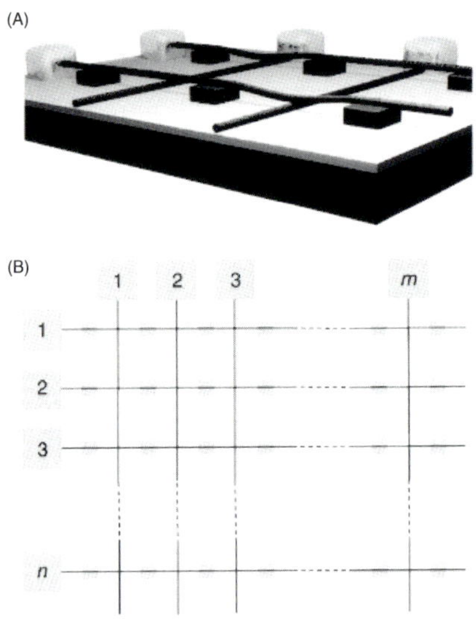

Schematics of freestanding nanotube device architecture with multiplex addressing (A). Threedimensional view of a suspended crossbar array showing four junctions with two elements in the ON (contact) state and two elements in the OFF (separated) state. (B) Top view of an n × m device array

The concept of the bitable device was demonstrated by current-voltage (I–V) behaviors of suspended, crossed nanotube devices made from SWNT ropes (~50 nm in diameter), with junction gap ~150 nm by mechanical manipulation under an optical microscope. The resistances of the OFF state of the devices were found consistently 10-fold larger than the ON state.

In the integrated system, electrical contacts are made only at one end of each of the lower and upper sets of nanoscale wires in the crossbar array, and thus, many device elements can be addressed from a limited number of contacts. This approach suggests a highly integrated, fast, and macroscopically addressable NRAM structure that could overcome the fundamental limitations of semiconductor random access memory in size, speed, and cost. Integration levels as high as 1 × 1012 elements per square centimeter and switching time down to ~5ps (200-GHz operation frequency) using 5-nm device elements and 5-nm supports are envisioned while maintaining the addressability of many devices through the long (~10-m) SWNT wires. However, such small dimensions, in particular, the junction gap size, impose significant challenges in the nanofabrication of parallel device arrays.

Nanotweezers There are two types of carbon nanotube–based nanotweezers reported by Kim and Lieber in 1999 and Akita et al. in 2001, respectively. Both nanotweezers employ MWNTs as tweezers' arms that are actuated by electrostatic forces. The applications of these nanotweezers include the manipulation of nanostructures and two-tip STM or atomic force microscope (AFM) probes.

The fabrication process of the carbon nanotube–based nanotweezers reported by Kim and Lieber is illustrated in figure. Freestanding electrically independent electrodes were deposited onto tapered glass micropipettes with end diameters of 100 nm. MWNT or SWNT bundles with diameters 20–50 nm were attached to the two gold electrodes, under the direct view of an optical microscope operated in dark-field mode, using an adhesive. A SEM image of fabricated nanotube tweezers is shown in figure.

The electromechanical response of nanotube nanotweezers was investigated by applying bias voltages to the electrodes while simultaneously imaging the nanotube displacements under an optical microscope in dark-field mode. As the bias voltage increased from 0 to 8.3 V, the free ends of the tweezers' arms bent closer to each other from their relaxed position (at 0 V). The tweezers' arms relaxed to the original position when the applied voltage was removed, and this process could be repeated more than 10 times, producing the same displacement each time within the optical microscope resolution limit. These results demonstrated that the mechanical response was elastic and thus that neither the nanotubes nor the nanotube-electrode junctions deform inelastically. At 8.3 V, the distance between the tweezers' ends decreased by about 50% of the initial value, and as the voltage was increased further to 8.5V, the tweezers' arms suddenly closed.

The nanotube nanotweezers have been demonstrated successfully to manipulate nano-structures, such as fluorescently labeled polystyrene spheres and -SiC nanocluster and GaAs nanowires.

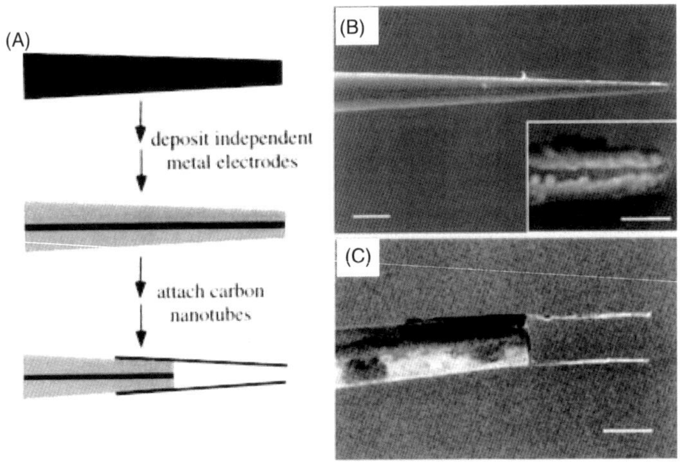

Overview of the fabrication process of carbon nanotube nanotweezers. (A) Schematic illustrating the deposition of two independent metal electrodes and the subsequent attachment of carbon nanotubes to these electrodes. (B) SEM image of the end of a tapered glass structure after the two deposition steps. Scale bar, 1 m. The higher-resolution inset shows clearly that the electrodes are separated. Scale bar, 200 nm. (C) SEM image of nanotweezers after mounting two MWNT bundles on each electrode. Scale bar, 2 m.

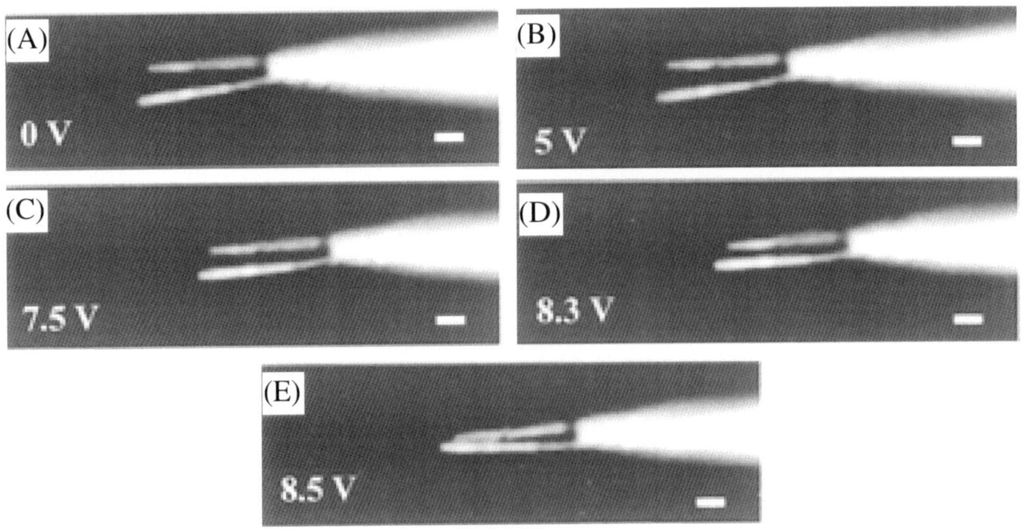

Dark-field-optical micrographs of electromechanical response of the nanotube nanotweezers with respect to different applied voltages. The scale bar is 1 m.

The carbon nanotube–based nanotweezers, reported by Akita et al., are shown in figure. Commercially available Si AFM cantilevers were employed as the device body. A Ti/Pt film was coated on the tip of the cantilever and connected to three Al interconnects that were patterned on the cantilever by a conventional lithographic technique as shown in figure. The Ti/Pt film was separated into two by focused ion beam etching. These two parts were independently connected to Al interconnects as shown in figure. DC voltage was applied between the separated Ti/Pt tips, through the Al interconnects, to operate the tweezers after attaching two arms of nanotubes on them.

The attachment of the nanotubes was carried out in a specially designed field-emission-type SEM with three independent movable stages. The Si cantilever and the nanotube cartridge were mounted on two different stages. A third stage, where a tungsten needle was installed, was used for the fine adjustment of the position of the nanotubes after being mounted to the Si tip. When the metal-coated Si tip was manipulated to be in contact with a target nanotube, an amorphous carbon film was deposited on this contact portion by the electron-beam dissociation of contaminants, mainly hydrocarbons, in the SEM chamber. The target nanotube was finally pulled away from the cartridge. Another nanotube was also attached in the same manner. The position of the two nanotube arms was adjusted to be parallel by using the stage with the tungsten needle and fixed by deposition of the carbon films at the base of the arms. The nanotube arms were also coated with an ultrathin carbon film (a few to several nanometers) to achieve insulation from the outside. This film coating prevents large current flows when the two nanotube arms close or pick up a conductive particle. Figure shows a SEM image of a typical pair of nanotube nanotweezers prepared this way. Two arms of the nanotubes were fixed at the most appropriate position on the Si tips. Their length was 2.5 m, and the separation between their tips was 780 nm.

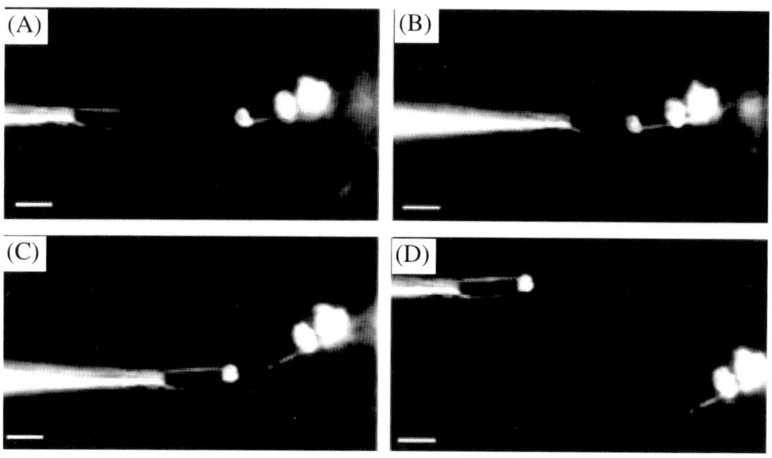

Dark-field-optical micrographs showing the sequential process of nanotweezer manipulation of polystyrene nanoclusters containing fluorescent dye molecules.

The operation of the nanotube nanotweezers was examined by in situ SEM. Various voltages were applied between the two arms to get them to close because of the electrostatic attraction force. Figures 11(b)–7(d) show the motion of the nanotube arms as a function of the applied voltage *V*. It is clearly seen that the arms bent and the separation between the tips decreased with increasing applied voltage. The separation became 500 nm at $V = 4$ V and zero at $V > 4$ 5V. It is noted that the motion in Figs. 11(a)–11(d) could be repeated many times without any permanent deformation, showing that carbon nanotubes are ideal materials for building NEMS.

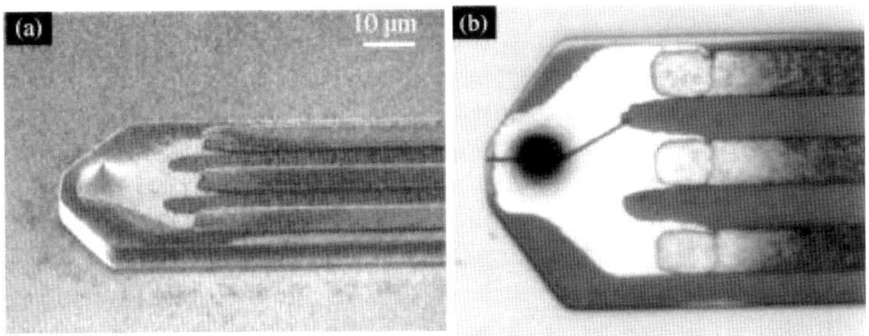

SEM images of a Si cantilever as a base for nanotube nanotweezers. (a) A Ti/Pt film was coated on the tip and connected to three Al lines patterned on the cantilever. (b) The Ti/Pt film was separated into two by a focused ion beam, and these two were connected to the one and two Al lines, respectively.

Rotational Motors A carbon nanotube–based rotational motor, reported by Fennimore et al. in 2003, is conceptually illustrated in Fig. The rotational element (R), a solid rectangular metal plate serving as a rotor, is attached transversely to a suspended support shaft. The support shaft ends are embedded in electrically conducting anchors (A1, A2) that rest on the oxidized surface of a silicon chip. The rotor plate assembly is surrounded by three fixed stator electrodes: two “in-plane” stators (S1, S2) are horizontally opposed and rest on the silicon oxide surface, and the third “gate” stator (S3) is buried beneath the surface. Four independent (DC and/or appropriately phased AC) voltage signals, one to the rotor plate and three to the stators, are applied to control the position, speed, and direction of rotation of the rotor plate. The key component in the assembly is a MWNT, which serves simultaneously as the rotor plate support shaft and the electrical feedthrough to the rotor plate; most importantly, it is also the source of rotational freedom.

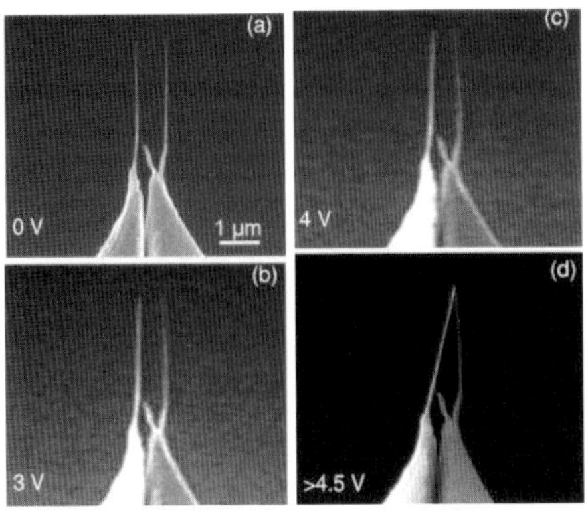

SEM images of the motion of nanotube arms in a pair of nanotweezers as a function of the applied voltage

Integrated synthetic nanoelectromechanical systems (NEMS) actuator. (a) Conceptual drawing of nanoactuator. (b) SEM image of nanoactuator just prior to HF etching. Scale bar, 300 nm

The nanoactuator was constructed using lithographic methods. MWNTs in suspension were deposited on a doped silicon substrate covered with 1 m of SiO_2. The nanotubes were located using an AFM or a SEM. The remaining actuator components (in-plane rotor plate, in-plane stators, anchors, and electrical leads) were then patterned using electronbeam lithography. An HF etch was used to remove roughly 500 nm of the SiO_2 surface to provide clearance for the rotor plate. The conducting Si substrate here serves as the gate stator. Figure shows an actuator device prior to etching. Typical rotor plate dimensions were 250–500 nm on a side.

The performance of the nanoactuator was examined in situ inside the SEM chamber. Visible rotation could be obtained by applying DC voltages up to 50 V between the rotor plate and the gate stator. When the applied voltage was removed, the rotor plate would rapidly return to its original horizontal position. To exploit the intrinsic low-friction-bearing behavior afforded by the perfectly nested shells of MWNTs, the MWNT supporting shaft was modified in situ by successive application of very large stator voltages. The process resulted in fatigue and eventually shear failure of the outer nanotube shells. In the "free" state, the rotor plate was still held in position axially by the intact nanotube core shells but could be azimuthally positioned, using an appropriate combination of stator signals, to any arbitrary angle between 0° and 360°. Figure 13 shows a series of still SEM images, recorded from an actuated device in the free state, being "walked" through one complete rotor plate revolution using quasi-static DC stator voltages. The stator voltages were adjusted sequentially to attract the rotor plate edge to successive stators. By reversing the stator voltage sequence, the rotor plate rotation could be reversed in an equally controlled fashion. Finite frequency operation of the actuator was also performed using a variety of suitably phased AC and DC voltage signals to the three stators and rotor plate. The rotor plate was successfully flipped between the extreme horizontal (90° and 270°) positions. The experiments show that the MWNT clearly serves as a reliable, presumably wear-free, NEMS element providing rotational freedom. No apparent wear or degradation in performance was observed after many thousands of cycles of rotations.

The potential applications of the MWNT-based actuators include ultra-high-density optical sweeping and switching elements, paddles for inducing and/or detecting fluid motion in microfluidics systems, gated catalysts in wet chemistry reactions, biomechanical elements in biological systems, or general (potentially chemically functionalized) sensor elements.

Nanoelectromechanical Relay

Nano-Electro-Mechanical (NEM) relays are nano-scale switches that can be mechanically actuated by an electrical signal. Unlike conventional CMOS transistors, NEM relays exhibit zero off-state leakage and very sharp on-off transitions. As a result, NEM relays can be potentially used to design highly energy-efficient digital systems. NEM relays are also excellent candidates for programmable routing switches in Field Programmable Gate Arrays (FPGAs) due to their potentially low on-state resistances despite their long mechanical delays. Low-temperature fabrication of NEM relays creates opportunities for their integration on top of silicon CMOS circuits. Hysteresis properties of NEM relays can enable their use as FPGA programmable routing switches without requiring additional routing SRAM cells.

Operation

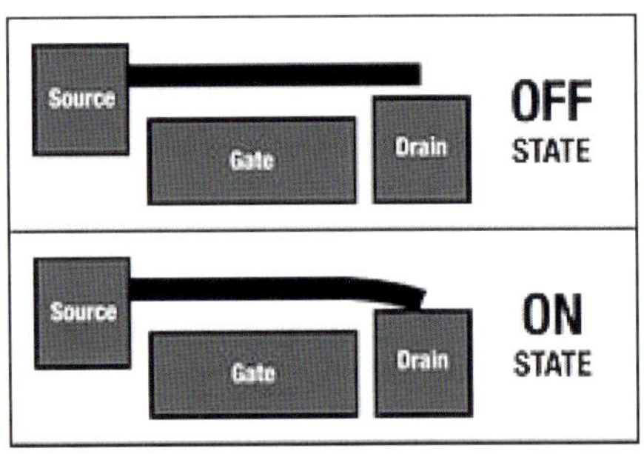

Schematic of a three terminal electromechanical relay

A NEM relay can be fabricated in two, three, or four terminal configurations. A three terminal relay is composed of a source (input), drain (output), and a gate (actuation terminal). Attached to the source is a cantilevered beam that can be bent into contact with the drain in order to make an electrical connection. When a significant voltage differential is applied between the beam and gate, and the electrostatic force overcomes the elastic force of the beam enough to bend it into contact with the drain, the device "pulls in" and forms an electrical connection. In the off position, the source and drain are separated by an air gap. This physical separation allows NEM relays to have zero current leakage, and very sharp on/off transitions.

The nonlinear nature of the electric field, and adhesion between the beam and drain cause the device to "pull out" and lose connection at a lower voltage than the voltage at which it pulls in. This hysteresis effect means there is a voltage between the pull in voltage, and the pull out voltage that will not change the state of the relay, no matter what its initial state is. This property is very useful in applications where information needs to be stored in the circuit, such as in static random-access memory.

Fabrication

NEM relays are usually fabricated using surface micromachining techniques typical of microelectromechanical systems (MEMS). Laterally actuated relays are constructed by first depositing two or more layers of material on a silicon wafer. The upper structural layer is photolithographically patterned in order to form isolated blocks of the uppermost material. The layer below is then selectively etched away, leaving thin structures, such as the relay's beam, cantilevered above the wafer, and free to bend laterally. A common set of materials used in this process is polysilicon as the upper structural layer, and silicon dioxide as the sacrificial lower layer.

NEM relays can be fabricated using a back end of line compatible process, allowing them to be built on top of CMOS. This property allows NEM relays to be used to significantly reduce the area of certain circuits. For example, a CMOS-NEM relay hybrid inverter occupies 0.03 μm^2, one-third the area of a 45 nm CMOS inverter.

Applications

- Mechanical computing

Due to transistor leakage, there is a limit to the theoretical efficiency of CMOS logic. This efficiency barrier ultimately prevents continued increases in computing power in power-constrained applications. While NEM relays have significant switching delays, their small size and fast switching speed when compared to other relays means that mechanical computing utilizing NEM Relays could prove a viable replacement for typical CMOS based integrated circuits, and break this CMOS efficiency barrier.

A NEM relay switches mechanically about 1000 times slower than a solid-state transistor takes to switch electrically. While this makes using NEM relays for computing a significant challenge, their low resistance would allow many NEM relays to be chained together and switch all at once, performing a single large calculation. On the other hand, transistor logic has to be implemented in small cycles of calculations, because their high resistance does not allow many transistors to be chained together while maintaining signal integrity. Therefore, it would be possible to create a mechanical computer using NEM relays that operates at a much lower clock speed than CMOS logic, but performs larger, more complex calculations during each cycle. This would allow a NEM relay based logic to perform to standards comparable to current CMOS logic.

There are many applications, such as in the automotive, aerospace, or geothermal exploration businesses, in which it would be beneficial to have a microcontroller that could operate at very high temperatures. However, at high temperatures, semiconductors used in typical microcontrollers begin to fail as the electrical properties of the materials they are made of degrade, and the transistors no longer function. NEM relays do not rely on the electrical properties of materials to actuate, so a mechanical computer utilizing NEM relays would be able to operate in such conditions. NEM relays have been successfully tested at up to 500°C, but could theoretically withstand much higher temperatures.

- Field-programmable gate arrays

The zero leakage current, low energy usage, and ability to be layered on top of CMOS properties of NEM relays make them a promising candidate for usage as routing switches in field-programmable gate arrays (FPGA). A FPGA utilizing a NEM relay to replace each routing switch and its corresponding static random-access memory block could allow for a significant reduction in programming delay, power leakage, and chip area compared to a typical 22nm CMOS based FPGA. This area reduction mainly comes from the fact that the NEM relay routing layer can be built on top of the CMOS layer of the FPGA.

Nanotechnology

Nanotechnology is the manipulation and manufacture of materials and devices on the scale of atoms or small groups of atoms. The "nanoscale" is typically measured in nanometres, or billionths of a metre (*nanos*, the Greek word for "dwarf," being the source of the prefix), and materials built at this scale often exhibit distinctive physical and chemical properties due to quantum mechanical effects. Although usable devices this small may be decades away, techniques for working at the nanoscale have become essential to electronic engineering, and nanoengineered materials have begun to appear in consumer products. For example, billions of microscopic "nanowhiskers," each about 10 nanometres in length, have been molecularly hooked onto natural and synthetic fibres to impart stain resistance to clothing and other fabrics; zinc oxide nanocrystals have been used to create invisible sunscreens that block ultraviolet light; and silver nanocrystals have been embedded in bandages to kill bacteria and prevent infection.

Possibilities for the future are numerous. Nanotechnology may make it possible to manufacture lighter, stronger, and programmable materials that require less energy to produce than conventional materials, that produce less waste than with conventional manufacturing, and that promise greater fuel efficiency in land transportation, ships, aircraft, and space vehicles. Nanocoatings for both opaque and translucent surfaces may render them resistant to corrosion, scratches, and radiation. Nanoscale electronic, magnetic, and mechanical devices and systems with unprecedented levels of information processing may be fabricated, as may chemical, photochemical, and biological sensors for protection, health care, manufacturing, and the environment; new photoelectric materials that will enable the manufacture of cost-efficient solar-energy panels; and molecular-semiconductor hybrid devices that may become engines for the next revolution in the information age. The potential for improvements in health, safety, quality of life, and conservation of the environment are vast.

At the same time, significant challenges must be overcome for the benefits of nanotechnology to be realized. Scientists must learn how to manipulate and characterize individual atoms and small groups of atoms reliably. New and improved tools are needed to control the properties and structure of materials at the nanoscale; significant improvements in computer simulations of atomic and molecular structures are essential to the understanding of this realm. Next, new tools and approaches are needed for assembling atoms and molecules into nanoscale systems and for the further assembly of small systems into more-complex objects. Furthermore, nanotechnology products must provide not only improved performance but also lower cost. Finally, without integration of nanoscale objects with systems at the micro- and macroscale (that is, from millionths of a metre up to the millimetre scale), it will be very difficult to exploit many of the unique properties found at the nanoscale.

Nanotechnology is highly interdisciplinary, involving physics, chemistry, biology, materials science, and the full range of the engineering disciplines. The word *nanotechnology* is widely used as shorthand to refer to both the science and the technology of this emerging field. Narrowly defined, nanoscience concerns a basic understanding of physical, chemical, and biological properties on atomic and near-atomic scales. Nanotechnology, narrowly defined, employs controlled manipulation of these properties to create materials and functional systems with unique capabilities.

In contrast to recent engineering efforts, nature developed "nanotechnologies" over billions of years, employing enzymes and catalysts to organize with exquisite precision different kinds of atoms and molecules into complex microscopic structures that make life possible. These natural products are built with great efficiency and have impressive capabilities, such as the power to harvest solar energy, to convert minerals and water into living cells, to store and process massive amounts of data using large arrays of nerve cells, and to replicate perfectly billions of bits of information stored in molecules of deoxyribonucleic acid (DNA).

There are two principal reasons for qualitative differences in material behaviour at the nanoscale (traditionally defined as less than 100 nanometres). First, quantum mechanical effects come into play at very small dimensions and lead to new physics and chemistry. Second, a defining feature at the nanoscale is the very large surface-to-volume ratio of these structures. This means that no atom is very far from a surface or interface, and the behaviour of atoms at these higher-energy sites have a significant influence on the properties of the material. For example, the reactivity of a metal catalyst particle generally increases appreciably as its size is reduced—macroscopic gold is chemically inert, whereas at nanoscales gold becomes extremely reactive and catalytic and even melts at a lower temperature. Thus, at nanoscale dimensions material properties depend on and change with size, as well as composition and structure.

Using the processes of nanotechnology, basic industrial production may veer dramatically from the course followed by steel plants and chemical factories of the past. Raw materials will come from the atoms of abundant elements—carbon, hydrogen, and silicon—and these will be manipulated into precise configurations to create nanostructured materials that exhibit exactly the right properties for each particular application. For example, carbon atoms can be bonded together in a number of different geometries to create variously a fibre, a tube, a molecular coating, or a wire, all with the superior strength-to-weight ratio of another carbon material—diamond. Additionally, such material processing need not require smokestacks, power-hungry industrial machinery, or intensive human labour. Instead, it may be accomplished either by "growing" new structures through some combination of chemical catalysts and synthetic enzymes or by building them through new techniques based on patterning and self-assembly of nanoscale materials into useful predetermined designs. Nanotechnology ultimately may allow people to fabricate almost any type of material or product allowable under the laws of physics and chemistry. While such possibilities seem remote, even approaching nature's virtuosity in energy-efficient fabrication would be revolutionary.

Even more revolutionary would be the fabrication of nanoscale machines and devices for incorporation into micro- and macroscale systems. Once again, nature has led the way with the fabrication of both linear and rotary molecular motors. These biological machines carry out such tasks as muscle contraction (in organisms ranging from clams to humans) and shuttling little packets of material around within cells while being powered by the recyclable, energy-efficient fuel adenosine triphosphate. Scientists are only beginning to develop the tools to fabricate functioning systems at such small scales, with most advances based on electronic or magnetic information processing and storage systems. The energy-efficient, reconfigurable, and self-repairing aspects of biological systems are just becoming understood.

The potential impact of nanotechnology processes, machines, and products is expected to be far-reaching, affecting nearly every conceivable information technology, energy source, agricultural product, medical device, pharmaceutical, and material used in manufacturing. Meanwhile, the dimensions of electronic circuits on semiconductors continue to shrink, with minimum feature sizes now reaching the nanorealm, under 100 nanometres. Likewise, magnetic memory materials, which form the basis of hard disk drives, have achieved dramatically greater memory density as a result of nanoscale structuring to exploit new magnetic effects at nanodimensions. These latter two areas represent another major trend, the evolution of critical elements of microtechnology into the realm of nanotechnology to enhance performance. They are immense markets driven by the rapid advance of information technology.

Feynman was intrigued by biology and pointed out that cells are very tiny, but they are very active; they manufacture various substances; they walk around; they wiggle; and they do all kinds of marvelous things—all on a very small scale. Also, they store information. Consider the possibility that we too can make a thing very small which does what we want—that we can manufacture an object that maneuvers at that level.

He also considered using big tools to make smaller tools that could make yet smaller tools, eventually obtaining nanoscale tools for directly manipulating atoms and molecules. In considering what all this might mean, Feynman declared.

Perhaps the biggest barrier to following these prophetic thoughts was simply the immediate lack of tools to manipulate and visualize matter at such a small scale. The availability of tools has always been an enabling aspect of the advance of all science and technology.

Starting with a 1981 paper in the and following with two popular books, American scientist K. Eric Drexler became one of the foremost advocates of nanotechnology. In fact, Drexler was the first person anywhere to receive a Ph.D. in molecular nanotechnology (from the Massachusetts Institute of Technology). In his written works he takes a molecular view of the world and envisions molecular machines doing much of the work of the future. For example, he refers to "assemblers," which will manipulate individual atoms to manufacture structures, and "replicators," which will be able to make multiple copies of themselves in order to save time dealing with the billions of atoms needed to make objects of useful size. Drexler wrote:

Cells and tissues in the human body are built and maintained by molecular machinery, but sometimes that machinery proves inadequate: viruses multiply, cancer cells spread, or systems age and deteriorate. As one might expect, new molecular machines and computers of subcellular size could support the body's own mechanisms. Devices containing nanocomputers interfaced to molecular sensors and effectors could serve as an augmented immune system, searching out and destroying viruses and cancer cells. Similar devices programmed as repair machines could enter living cells to edit out viral DNA sequences and repair molecular damage. Such machines would bring surgical control to the molecular level, opening broad new horizons in medicine.

Drexler's futurist visions have stimulated much thought, but the assembler approach has failed to account for the strong influence of atomic and molecular forces (i.e., the chemistry) at such dimensions. The controversy surrounding these popularizations, and the potential dangers of entities such as intelligent replicators (however remote), have stimulated debate over the ethical and societal implications of nanotechnology.

A number of key technological milestones have been achieved by working pioneers. Molecular beam epitaxy, invented by Alfred Cho and John Arthur at Bell Labs in 1968 and developed in the 1970s, enabled the controlled deposition of single atomic layers. This tool provided for nanostructuring in one dimension as atomic layers were grown one upon the next. It subsequently became important in the area of compound semiconductor device fabrication. For example, sandwiching one-nanometre-thick layers of nonmagnetic-sensor materials between magnetic layers in computer disk drives resulted in large increases in storage capacity, and a similar use of nanostructuring resulted in more energy-efficient semiconductor lasers for use in compact disc players.

In 1981 Gerd Binnig and Heinrich Rohrer developed the scanning tunneling microscope at IBM's laboratories in Switzerland. This tool provided a revolutionary advance by enabling scientists to image the position of individual atoms on surfaces. It earned Binnig and Rohrer a Nobel Prize in 1986 and spawned a wide variety of scanning probe tools for nanoscale observations.

The observation of new carbon structures marked another important milestone in the advance of nanotechnology, with Nobel Prizes for the discoverers. In 1985 Robert F. Curl, Jr., Harold W. Kroto, and Richard E. Smalley discovered the first fullerene, the third known form of pure carbon (after diamond and graphite). They named their discovery buckminsterfullerene ("buckyball") for its resemblance to the geodesic domes promoted by the American architect R. Buckminster Fuller. Technically called C_{60} for the 60 carbon atoms that form their hollow spherical structure, buckyballs resemble a football one nanometre in diameter. In 1991 Sumio Iijima of NEC Corporation in Japan discovered carbon nanotubes, in which the carbon ringlike structures are extended from spheres into long tubes of varying diameter. Taken together, these new structures surprised and excited the imaginations of scientists about the possibilities of forming well-defined nanostructures with unexpected new properties.

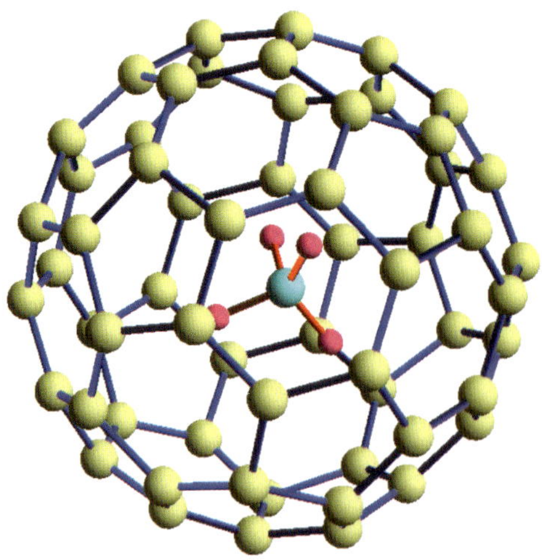

The structure of buckminsterfullerene (C_{60})

The scanning tunneling microscope not only allowed for the imaging of atoms by scanning a sharp probe tip over a surface, but it also allowed atoms to be "pushed" around on the surface. With a slight bias voltage applied to the probe tip, certain atoms could be made to adhere to the tip used for imaging and then to be released from it. Thus, in 1990 Donald Eigler spelled out the letters of his company's logo, IBM, by moving 35 xenon atoms into place on a nickel surface. This demonstration caught the public's attention because it showed the precision of the emerging nanoscale tools.

Properties at the Nanoscale

At nanoscale dimensions the properties of materials no longer depend solely on composition and structure in the usual sense. Nanomaterials display new phenomena associated with quantized effects and with the preponderance of surfaces and interfaces.

Quantized effects arise in the nanometre regime because the overall dimensions of objects are comparable to the characteristic wavelength for fundamental excitations in materials. For example, electron wave functions (*see also* de Broglie wave) in semiconductors are typically on the order of 10 to 100 nanometres. Such excitations include the wavelength of electrons, photons, phonons, and magnons, to name a few. These excitations carry the quanta of energy through materials and thus determine the dynamics of their propagation and transformation from one form to another. When the size of structures is comparable to the quanta themselves, it influences how these excitations move through and interact in the material. Small structures may limit flow, create wave interference effects, and otherwise bring into play quantum mechanical selection rules not apparent at larger dimensions.

Electronic and Photonic Behaviour

Quantum mechanical properties for confinement of electrons in one dimension have long been exploited in solid-state electronics. Semiconductor devices are grown with thin layers of differing composition so that electrons (or "holes" in the case of missing electron charges) can be confined in specific regions of the structure (known as quantum wells). Thin layers with larger energy bandgaps can serve as barriers that restrict the flow of charges to certain conditions under which they can "tunnel" through these barriers—the basis of resonant tunneling diodes. Superlattices are periodic structures of repeating wells that set up a new set of selection rules which affect the conditions for charges to flow through the structure. Superlattices have been exploited in cascade lasers to achieve far infrared wavelengths. Modern telecommunications is based on semiconductor lasers that exploit the unique properties of quantum wells to achieve specific wavelengths and high efficiency.

The propagation of photons is altered dramatically when the size and periodicity of the transient structure approach the wavelength of visible light (400 to 800 nanometres). When photons propagate through a periodically varying dielectric constant—for example, semiconductor posts surrounded by air—quantum mechanical rules define and limit the propagation of the photons depending on their energy (wavelength). This new behaviour is analogous to the quantum mechanical rules that define the motion of electrons through crystals, giving bandgaps for semiconductors. In one dimension, compound semiconductor superlattices can be grown epitaxially with the alternating layers having different dielectric constants, thus providing highly reflective mirrors for specific wavelengths as determined by the repeat distance of layers in the superlattice. These structures are used to provide "built-in" mirrors for vertical-cavity surface-emitting lasers, which are used in communications applications. In two and three dimensions, periodic structures known as photonic crystals offer additional control over photon propagation.

Photonic crystals are being explored in a variety of materials and periodicities, such as two-dimensional hexagonal arrays of posts fabricated in compound semiconductors or stacked loglike arrays of silicon bars in three dimensions. The dimensions of these structures depend on the wavelength of light being propagated and are typically in the range of a few hundred nanometres for wavelengths in the visible and near infrared. Photonic crystal properties based on nanostructured materials offer the possibility of confining, steering, and separating light by wavelength on unprecedented small scales and of creating new devices such as lasers that require very low currents to initiate lasing (called near-thresholdless lasers). These structures are being extensively investigated as the tools for nanostructuring materials are steadily advancing. Researchers are particularly interested in the infrared wavelengths, where dimensional control is not as stringent as at the shorter visible wavelengths and where optical communications and chemical sensing provide motivation for potential new applications.

Magnetic, Mechanical, and Chemical Behaviour

Nanoscale materials also have size-dependent magnetic behaviour, mechanical properties, and chemical reactivity. At very small sizes (a few nanometres), magnetic nanoclusters have a single magnetic domain, and the strongly coupled magnetic spins on each atom combine to produce a particle with a single "giant" spin. For example, the giant spin of a ferromagnetic iron particle rotates freely at room temperature for diameters below about 16 nanometres, an effect termed superparamagnetism. Mechanical properties of nanostructured materials can reach exceptional strengths. As a specific example, the introduction of two-nanometre aluminum oxide precipitates into thin films of pure nickel results in yield strengths increasing from 0.15 to 5 gigapascals, which is more than twice that for a hard bearing steel. Another example of exceptional mechanical properties at the nanoscale is the carbon nanotube, which exhibits great strength and stiffness along its longitudinal axis.

The preponderance of surfaces is a major reason for the change in behaviour of materials at the nanoscale. Since up to half of all the atoms in nanoparticles are surface atoms, properties such as electrical transport are no longer determined by solid-state bulk phenomena. Likewise, the atoms in nanostructures have a higher average energy than atoms in larger structures, because of the large proportion of surface atoms. For example, catalytic materials have a greater chemical activity per atom of exposed surface as the catalyst is reduced in size at the nanoscale. Defects and impurities may be attracted to surfaces and interfaces, and interactions between particles at these small dimensions can depend on the structure and nature of chemical bonding at the surface. Molecular monolayers may be used to change or control surface properties and to mediate the interaction between nanoparticles.

Surfaces and their interactions with molecular structures are basic to all biology. The intersection of nanotechnology and biotechnology offers the possibility of achieving new functions and properties with nanostructured surfaces. In this surface- and interface-dominated regime, biology does an exquisite job of selectively controlling functions through a combination of structure and chemical forces. The transcription of information stored in genes and the selectivity of biochemical reactions based on chemical recognition of complex molecules are examples where interfaces play the key role in establishing nanoscale behaviour. Atomic forces and chemical bonds dominate at these dimensions, while macroscopic effects—such as convection, turbulence, and momentum (inertial forces)—are of little consequence.

Nanomaterials

Properties at the nanoscale, material properties—electrical, optical, magnetic, mechanical, and chemical—depend on their exact dimensions. This opens the way for development of new and improved materials through manipulation of their nanostructure. Hierarchical assemblies of nanoscale-engineered materials into larger structures, or their incorporation into devices, provide the basis for tailoring radically new materials and machines.

The often-cited abalone seashell provides a beautiful example of how the combination of a hard, brittle inorganic material with nanoscale structuring and a soft, "tough" organic material can produce a strong, durable nanocomposite—basically, these nanocomposites are made of calcium carbonate "bricks" held together by a glycoprotein "glue." New engineered materials are emerging—such as polymer-clay nanocomposites—that are not only strong and tough but also lightweight and easier to recycle than conventional reinforced plastics. Such improvements in structural materials are particularly important for the transportation industry, where reduced weight directly translates into improved fuel economy. Other improvements can increase safety or decrease the impact on the environment of fabrication and recycling. Further advances, such as truly smart materials that signal their impending failure or are even able to self-repair flaws, may be possible with composites of the future.

Sensors are central to almost all modern control systems. For example, multiple sensors are used in automobiles for such diverse tasks as engine management, emission control, security, safety, comfort, vehicle monitoring, and diagnostics. While such traditional applications for physical sensing generally rely on microscale sensing devices, the advent of nanoscale materials and structures has led to new electronic, photonic, and magnetic nanosensors, sometimes known as "smart dust." Because of their small size, nanosensors exhibit unprecedented speed and sensitivity, extending in some cases down to the detection of single molecules. For example, nanowires made of carbon nanotubes, silicon, or other semiconductor materials exhibit exceptional sensitivity to chemical species or biological agents. Electrical current through nanowires can be altered by having molecules attached to their surface that locally perturb their electronic band structure. By means of nanowire surfaces coated with sensor molecules that selectively attach particular species, charge-induced changes in current can be used to detect the presence of those species. This same strategy is adopted for many classes of sensing systems. New types of sensors with ultrahigh sensitivity and specificity will have many applications; for example, sensors that can detect cancerous tumours when they consist of only a few cells would be a very significant advance.

Nanomaterials also make excellent filters for trapping heavy metals and other pollutants from industrial wastewater. One of the greatest potential impacts of nanotechnology on the lives of the majority of people on Earth will be in the area of economical water desalination and purification. Nanomaterials will very likely find important use in fuel cells, bioconversion for energy, bioprocessing of food products, waste remediation, and pollution-control systems.

A recent concern regarding nanoparticles is whether their small sizes and novel properties may pose significant health or environmental risks. In general, ultrafine particles—such as the carbon in photocopier toners or in soot produced by combustion engines and factories—have adverse respiratory and cardiovascular effects on people and animals. Studies are under way to determine if specific nanoscale particles pose higher risks that may require special regulatory restrictions. Of particular concern are potential carcinogenic risks from inhaled particles and the possibility for very small nanoparticles to cross the blood-brain barrier to unknown effect. Nanomaterials currently receiving attention from health officials include carbon nanotubes, buckyballs, and cadmium selenide quantum dots. Studies of the absorption through the skin of titanium oxide nanoparticles (used in sunscreens) are also planned. More far-ranging studies of the toxicity, transport, and overall fate of nanoparticles in ecosystems and the environment have not yet been undertaken. Some early animal studies, involving the introduction of very high levels of nanoparticles which resulted in the rapid death of many of the subjects, are quite controversial.

Biomedicine and Health Care

Drug Delivery

Nanotechnology promises to impact medical treatment in multiple ways. First, advances in nanoscale particle design and fabrication provide new options for drug delivery and drug therapies. More than half of the new drugs developed each year are not water-soluble, which makes their delivery difficult. In the form of nanosized particles, however, these drugs are more readily transported to their destination, and they can be delivered in the conventional form of pills.

More important, nanotechnology may enable drugs to be delivered to precisely the right location in the body and to release drug doses on a predetermined schedule for optimal treatment. The general approach is to attach the drug to a nanosized carrier that will release the medicine in the body over an extended period of time or when specifically triggered to do so. In addition, the surfaces of these nanoscale carriers may be treated to seek out and become localized at a disease site—for example, attaching to cancerous tumours. One type of molecule of special interest for these applications is an organic dendrimer. A dendrimer is a special class of polymeric molecule that weaves in and out from a hollow central region. These spherical "fuzz balls" are about the size of a typical protein but cannot unfold like proteins. Interest in dendrimers derives from the ability to tailor their cavity sizes and chemical properties to hold different therapeutic agents. Researchers hope to design different dendrimers that can swell and release their drug on exposure to specifically recognized molecules that indicate a disease target. This same general approach to nanoparticle-directed drug delivery is being explored for other types of nanoparticles as well.

Another approach involves gold-coated nanoshells whose size can be adjusted to absorb light energy at different wavelengths. In particular, infrared light will pass through several centimetres of body tissue, allowing a delicate and precise heating of such capsules in order to release the therapeutic substance within. Furthermore, antibodies may be attached to the outer gold surface of the shells to cause them to bind specifically to certain tumour cells, thereby reducing the damage to surrounding healthy cells.

Bioassays

A second area of intense study in nanomedicine is that of developing new diagnostic tools. Motivation for this work ranges from fundamental biomedical research at the level of single genes or cells to point-of-care applications for health delivery services. With advances in molecular biology, much diagnostic work now focuses on detecting specific biological "signatures." These analyses are referred to as bioassays. Examples include studies to determine which genes are active in response to a particular disease or drug therapy. A general approach involves attaching fluorescing dye molecules to the target biomolecules in order to reveal their concentration.

Another approach to bioassays uses semiconductor nanoparticles, such as cadmium selenide, which emit light of a specific wavelength depending on their size. Different-size particles can be tagged to different receptors so that a wider variety of distinct colour tags are available than can be distinguished for dye molecules. The degradation in fluorescence with repeated excitation for dyes is avoided. Furthermore, various-size particles can be encapsulated in latex beads and their resulting wavelengths read like a bar code. This approach, while still in the exploratory stage, would allow for an enormous number of distinct labels for bioassays.

Another nanotechnology variation on bioassays is to attach one half of the single-stranded complementary DNA segment for the genetic sequence to be detected to one set of gold particles and the other half to a second set of gold particles. When the material of interest is present in a solution, the two attachments cause the gold balls to agglomerate, providing a large change in optical properties that can be seen in the colour of the solution. If both halves of the sequence do not match, no agglomeration will occur and no change will be observed.

Approaches that do not involve optical detection techniques are also being explored with nanoparticles. For example, magnetic nanoparticles can be attached to antibodies that in turn recognize and attach to specific biomolecules. The magnetic particles then act as tags and "handlebars" through which magnetic fields can be used for mixing, extracting, or identifying the attached biomolecules within microlitre- or nanolitre-sized samples. For example, magnetic nanoparticles stay magnetized as a single domain for a significant period, which enables them to be aligned and detected in a magnetic field. In particular, attached antibody–magnetic-nanoparticle combinations rotate slowly and give a distinctive magnetic signal. In contrast, magnetically tagged antibodies that are not attached to the biological material being detected rotate more rapidly and so do not give the same distinctive signal.

Microfluidic systems, or "labs-on-chips," have been developed for biochemical assays of minuscule samples. Typically cramming numerous electronic and mechanical components into a portable unit no larger than a credit card, they are especially useful for conducting rapid analysis in the field. While these microfluidic systems primarily operate at the microscale (that is, millionths of a metre), nanotechnology has contributed new concepts and will likely play an increasing role in the future. For example, separation of DNA is sensitive to entropic effects, such as the entropy required to unfold DNA of a given length. A new approach to separating DNA could take advantage of its passage through a nanoscale array of posts or channels such that DNA molecules of different lengths would uncoil at different rates.

Other researchers have focused on detecting signal changes as nanometre-wide DNA strands are threaded through a nanoscale pore. Early studies used pores punched in membranes by viruses; artificially fabricated nanopores are also being tested. By applying an electric potential across the membrane in a liquid cell to pull the DNA through, changes in ion current can be measured as different repeating base units of the molecule pass through the pores. Nanotechnology-enabled advances in the entire area of bioassays will clearly impact health care in many ways, from early detection, rapid clinical analysis, and home monitoring to new understanding of molecular biology and genetic-based treatments for fighting disease.

Assistive Devices and Tissue Engineering

Another biomedical application of nanotechnology involves assistive devices for people who have lost or lack certain natural capabilities. For example, researchers hope to design retinal implants for vision-impaired individuals. The concept is to implant chips with photodetector arrays to transmit signals from the retina to the brain via the optic nerve. Meaningful spatial information, even if only at a rudimentary level, would be of great assistance to the blind. Such research illustrates the tremendous challenge of designing hybrid systems that work at the interface between inorganic devices and biological systems.

Closely related research involves implanting nanoscale neural probes in brain tissue to activate and control motor functions. This requires effective and stable "wiring" of many electrodes to neurons. It is exciting because of the possibility of recovery of control for motor-impaired individuals. Studies employing neural stimulation of damaged spinal cords by electrical signals have demonstrated the return of some locomotion. Researchers are also seeking ways to assist in the regeneration and healing of bone, skin, and cartilage—for example, developing synthetic biocompatible or biodegradable structures with nanosized voids that would serve as templates for regenerating specific tissue while delivering chemicals to assist in the repair process. At a more sophisticated level, researchers hope to someday build nanoscale or microscale machines that can repair, assist, or replace more-complex organs.

Information Technology

Semiconductor experts agree that the ongoing shrinkage in "conventional" electronic devices will inevitably reach fundamental limits due to quantum effects such as "tunneling," in which electrons jump out of their prescribed circuit path and create atomic-scale interference between devices. At that point, radical new approaches to data storage and information processing will be required for further advances. For example, radically new systems have been imagined that are based on quantum computing or biomolecular computing.

Molecular Electronics

The use of molecules for electronic devices was suggested by Mark Ratner of Northwestern University and Avi Aviram of IBM as early as the 1970s, but proper nanotechnology tools did not become available until the turn of the 21st century. Wiring up molecules some half a nanometre wide and a few nanometres long remains a major challenge, and an understanding of electrical transport through single molecules is only beginning to emerge. A number of groups have been able to demonstrate molecular switches, for example, that could conceivably be used in computer memory or logic arrays. Current areas of research include mechanisms to guide the selection of molecules, architectures for assembling molecules into nanoscale gates, and three-terminal molecules for transistor-like behaviour. More-radical approaches include DNA computing, where single-stranded DNA on a silicon chip would encode all possible variable values and complementary strand interactions would be used for a parallel processing approach to finding solutions. An area related to molecular electronics is that of organic thin-film transistors and light emitters, which promise new applications such as video displays that can be rolled out like wallpaper and flexible electronic newspapers.

Nanotubes and Nanowires

Carbon nanotubes have remarkable electronic, mechanical, and chemical properties. Depending on their specific diameter and the bonding arrangement of their carbon atoms, nanotubes exhibit either metallic or semiconducting behaviour. Electrical conduction within a perfect nanotube is ballistic (negligible scattering), with low thermal dissipation. As a result, a wire made from a nanotube, or a nanowire, can carry much more current than an ordinary metal wire of comparable size. At 1.4 nanometres in diameter, nanotubes are about a hundred times smaller than the gate width of silicon semiconductor devices. In addition to nanowires for conduction, transistors, diodes, and simple logic circuits have been demonstrated by combining metallic and semiconductor carbon nanotubes. Similarly, silicon nanowires have been used to build experimental devices, such as field-effect transistors, bipolar transistors, inverters, light-emitting diodes, sensors, and even simple memory. A major challenge for nanowire circuits, as for molecular electronics, is connecting and integrating these devices into a workable high-density architecture. Ideally, the structure would be grown and assembled in place. Crossbar architectures that combine the function of wires and devices are of particular interest.

Single-electron Transistors

At nanoscale dimensions the energy required to add one additional electron to a "small island" (isolated physical region)—for example, through a tunneling barrier—becomes significant. This change in energy provides the basis for devising single-electron transistors. At low temperatures, where thermal fluctuations are small, various single-electron-device nanostructures are readily achievable, and extensive research has been carried out for structures with confined electron flow. However, room-temperature applications will require that sizes be reduced significantly, to the one-nanometre range, to achieve stable operation. For large-scale application with millions of devices, as found in current integrated circuits, the need for structures with very uniform size to maintain uniform device characteristics presents a significant challenge. Also, in this and many new nanodevices being explored, the lack of gain is a serious drawback limiting implementation in large-scale electronic circuits.

Spintronics

Spintronics refers to electronic devices that perform logic operations based on not just the electrical charge of carriers but also their spin. For example, information could be transported or stored through the spin-up or spin-down states of electrons. This is a new area of research, and issues include the injection of spin-polarized carriers, their transport, and their detection. The role of nanoscale structure and electronic properties of the ferromagnetic-semiconductor interface on the spin injection process, the growth of new ferromagnetic semiconductors with nanoscale control, and the possible use of nanostructured features to manipulate spin are all of interest.

Information Storage

Current approaches to information storage and retrieval include high-density, high-speed, solid-state electronic memories, as well as slower (but generally more spacious) magnetic and optical discs (*see* computer memory). As the minimum feature size for electronic processing approaches 100 nanometres, nanotechnology provides ways to decrease further the bit size of the stored information, thus increasing density and reducing interconnection distances for obtaining still-higher speeds. For example, the basis of the current generation of magnetic disks is the giant magnetoresistance effect. A magnetic read/write head stores bits of information by setting the direction of the magnetic field in nanometre-thick metallic layers that alternate between ferromagnetic and nonferromagnetic. Differences in spin-dependent scattering of electrons at the interface layers lead to resistance differences that can be read by the magnetic head. Mechanical properties, particularly tribology (friction and wear of moving surfaces), also play an important role in magnetic hard disk drives, since magnetic heads float only about 10 nanometres above spinning magnetic disks.

Another approach to information storage that is dependent on designing nanometre-thick magnetic layers is under commercial development. Known as magnetic random access memory (MRAM), a line of electrically switchable magnetic material is separated from a permanently magnetized layer by a nanoscale nonmagnetic interlayer. A resistance change that depends on the relative alignment of the fields is read electrically from a large array of wires through cross lines. MRAM will require a relatively small evolution from conventional semiconductor manufacturing, and it has the added benefit of producing nonvolatile memory (no power or batteries are needed to maintain stored memory states).

Still at an exploratory stage, studies of electrical conduction through molecules have generated interest in their possible use as memory. While still very speculative, molecular and nanowire approaches to memory are intriguing because of the small volume in which the bits of memory are stored and the effectiveness with which biological systems store large amounts of information.

Communications

Nanoscale structuring of optical devices, such as vertical-cavity surface-emitting lasers (VCSELs), quantum dot lasers, and photonic crystal materials, is leading to additional advances in communications technology.

VCSELs have nanoscale layers of compound semiconductors epitaxially grown into their structure—alternating dielectric layers as mirrors and quantum wells. Quantum wells allow the charge carriers to be confined in well-defined regions and provide the energy conversion into light at desired wavelengths. They are placed in the laser's cavity to confine carriers at the nodes of a standing wave and to tailor the band structure for more efficient radiative recombination. One-dimensional nanotechnology techniques involving precise growth of very thin epitaxial semiconductor layers were developed during the 1990s. Such nanostructuring has enhanced the efficiency of VCSELs and reduced the current required for lasing to start (called the threshold current). Because of improving performance and their compatibility with planar manufacturing technology, VCSELs are fast becoming a preferred laser source in a variety of communications applications.

More recently, the introduction of quantum dots (regions so small that they can be given a single electric charge) into semiconductor lasers has been investigated and found to give additional benefits—both further reductions in threshold current and narrower line widths. Quantum dots further confine the optical emission modes within a very narrow spectrum and give the lowest threshold current densities for lasing achieved to date in VCSELs. The quantum dots are introduced into the laser during the growth of strained layers, by a process called Stransky-Krastanov growth. They arise because of the lattice mismatch stress and surface tension of the growing film. Improvements in ways to control precisely the resulting quantum dots to a more uniform single size are still being sought.

Photonic crystals provide a new means to control the steering and manipulation of photons based on periodic dielectric lattices with repeat dimensions on the order of the wavelength of light. These materials can have very exotic properties, such as not allowing light within certain wavelengths to be propagated in a material based on the particular periodic structure. Photonic lattices can act as perfect wavelength-selective mirrors to reflect back incident light from all orientations. They provide the basis for optical switching, steering, and wavelength separation on unprecedented small scales. The periodic structures required for these artificial crystals can be configured as both two- and three-dimensional lattices. Optical sources, switches, and routers are being considered, with two-dimensional planar geometries receiving the most attention, because of their greater ease of fabrication.

Another potentially important communications application for nanotechnology is microelectromechanical systems (MEMS), devices sized at the micrometre level (millionths of a metre). MEMS are currently poised to have a major impact on communications via optical switching. In the future, electromechanical devices may shrink to nanodimensions to take advantage of the higher frequencies of mechanical vibration at smaller masses. The natural (resonant) frequency of vibration for small mechanical beams increases as their size decreases, so that little power is needed to drive them as oscillators. Their efficiency is rated by a quality factor, known as Q, which is a ratio of the energy stored per cycle versus the energy dissipated per cycle. The higher the Q, the more precise the absolute frequency of an oscillator. The Q is very high for micro- and nanoscale mechanical oscillators, and these devices can reach very high frequencies (up to microwave frequencies), making them potential low-power replacements for electronic-based oscillators and filters.

Mechanical oscillators have been made from silicon at dimensions of 10×100 nanometres, where more than 10 percent of the atoms are less than one atomic distance from the surface. While highly homogeneous materials can be made at these dimensions—for example, single-crystal silicon bars—surfaces play an increasing role at nanoscales, and energy losses increase, presumably because of surface defects and molecular species absorbed on surfaces.

It is possible to envision even higher frequencies, in what might be viewed as the ultimate in nanomechanical systems, by moving from nanomachined structures to molecular systems. As an example, multiwalled carbon nanotubes are being explored for their mechanical properties. When the ends of the outer nanotube are removed, the inner tube may be pulled partway out from the outer tube where van der Waals forces between the two tubes will supply a restoring force. The inner tube can thus oscillate, sliding back and forth inside the outer tube. The resonant frequency of oscillation for such structures is predicted to be above one gigahertz (one billion cycles per second). It is unknown whether connecting such systems to the macro world and protecting them from surface effects will ever be practical.

Micro-Opto-Electro-Mechanical Systems

Micromachines or systems that include optical components are often referred to as optical MEMS (o-MEMS), micro-opto-mechanical systems (MOMS), or micro-opto-electromechanical systems (MOEMS). Perhaps MOEMS is the most appropriate and general descriptor of these systems; it conveys the essential ideas about the size and nature of the elements that are integrated to form a system.

There are three primary characteristics that make MOEMS an important technology development: the first is the batch process by which the systems are fabricated; the second is the size of the elements in the systems; and the third, and perhaps most distinctive, is the possibility of endowing the optical elements in the system with precise and controllable motion. Movement of a micro-optical element permits dynamic manipulation of a light beam. This manipulation can involve (amplitude or wavelength) modulation, diffraction, reflection, refraction or simple spatial deflection. Any two or three of these operations can be combined to perform a complex operation on the light beam. The ability to carry out these operations, using miniaturized optical elements, is one of the key attributes that distinguishes MOEMS from classical physical optics.

The field of modem optics has been largely concerned with the generation, manipulation, guidance, or detection of light for information processing. The operation that is relevant to micro-opto-electro-mechanical-systems (MOEMS) is the manipulation of light in one, two or three dimensional space. Here, light is defined to be the electromagnetic radiation in the spectral band from about 200 nm to about 15 microns. This boundary definition is important because the wavelength of light that is manipulated or made to interact with micro-optical elements imposes a lower bound on the component size. This lower bound is a consequence of the laws of diffraction. In order to avoid unintentional diffraction effects, the feature sizes of micro-optical elements must be at least ten times larger than the wavelength of light that is intended to interact with the micro-optical element. If diffraction is the desired effect, then this restriction does not apply.

Conventional micromachines are comprised of micrometer-sized electrical and mechanical components integrated to form micro-electro-mechanical systems (MEMS). These systems are fabricated using the techniques and materials of microelectronics. The most common techniques are:

1. Bulk micromachining,
2. Wafer-to-wafer bonding,
3. Surface micromachining, and,
4. High-aspect ratio micromachining.

In bulk micromachining, a wet chemical etchant whose etching characteristics depend on the crystallographic surface chemistry of the substrate is used to selectively remove material from unmasked areas to define the geometry of the desired features. Wet chemical etching of this kind is generally anisotropic and a limited set of geometric features can be constructed in this way. To overcome this limitation, wafer-to-wafer bonding is used in conjunction with bulk micromachining to fuse together separately micromachined bulk wafers and achieve the desired geometric features. For further versatility in feature construction, surface micromachining is used. In this method, one starts with a substrate material which serves as a working surface. Multiple structural and sacrificial layers are deposited on it and then portions are selectively removed using a sequence of masking and etching steps. The etching is generally done using reactive ion etching-an isotropic etching process which is independent of the crystallographic surface. To fabricate thick (hundreds of microns to centimeters), high-aspect ratio structures, one uses deep UV lithography, in conjunction with reactive ion etching. In some cases, X-ray radiation from a synchrotron generator may also be used as the source for the lithography. Figures below show illustrations of two of the most commonly used methods for constructing micromachine features.

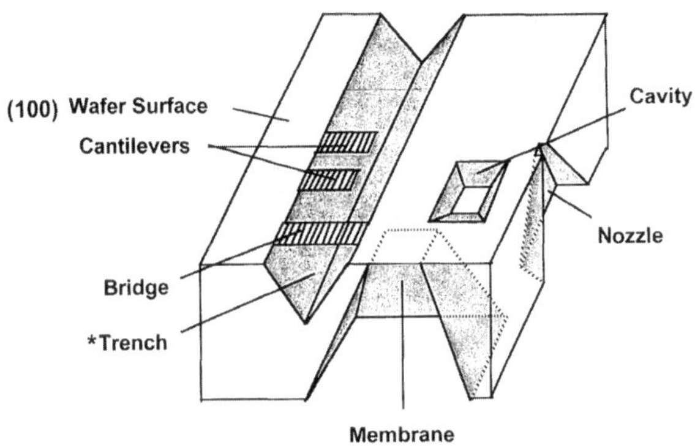

The processes described above have been extended to the construction of optical and fluidic components in both silicon and other substrate materials. The generality of the fabrication processes allows one to construct MEMS machines with a diversity of functionality. This functionality can be a result of a distinct class of features or a combination of classes. The major classes of features are:

> An arbitrary component with a composite of all common features and mechanical structures that can be etched in a piece of single-crystal silicon using bulk micromachining. Note that all etched walls are at the same angle as defined by the crystal orientation of the silicon. *Trench formed by the intersection of the (111) and (yyy) crystalline surfaces.

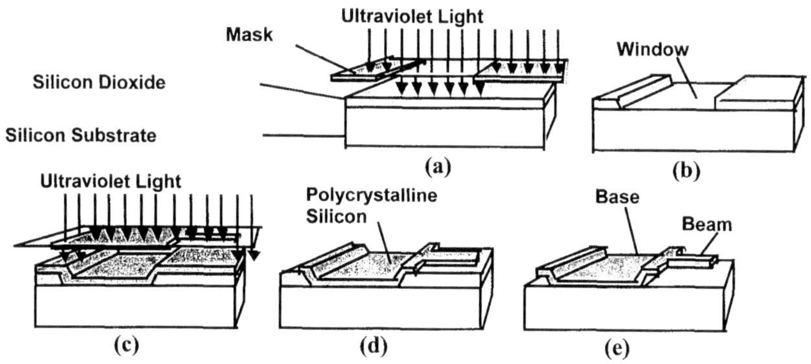

A single cycle in a common surface micromachining process. The process to build a single cantilever beam begins with the sacrificial material layer (silicon dioxide) being patterned and etched (a, b). Next, the structural material (polysilicon) is deposited over the entire surface. The polysilicon is then patterned and etched in the shape of the cantilever beam and base (c, d). Finally, the polysilicon is released by removing the remaining and underlying silicon dioxide (e).

- Micro-mechanics
  - replacement of passive lumped electrical elements with surface micromachined equivalents
  - micro-actuatable membranes
  - elements with micro-mechanical linear or rotary motion
- Micro-optics
  - diffractive, refractive and reflective micro-optical elements (fixed or movable) e.g., lenses, gratings, mirrors
  - micro-optical elements that exploit the free-space properties of light
  - self-aligned micro-optical elements

- Micro-fluidics
  - microchannels for fluid transport, storage, separation and reaction
  - micro-actuated valves for fluid control
  - micro-pumps for fluid movement

Each class of features can, and often does, include electronic devices that give the microsystem intelligence for control.

Radio Frequency Microelectromechanical System

Radio frequency (RF) microelectromechanical systems (MEMS) switches are devices that use mechanical movement to achieve a short circuit or an open circuit in an RF transmission.

Today, radio frequency microelectromechanical system (RF MEMS) technology is rapidly advancing, offering solutions to these challenges. The most widely recognised advantages are low loss, high isolation, near-perfect linearity, and large instantaneous bandwidth that conventional mechanical and semiconductor technologies fail to offer. Broadly, the applications in the field of defence fall in the categories of communication, phased array radars, and reconfigurable antennas. In addition to strategic sector applications, RF MEMS technological solutions are being sought in high volume in areas like automotive and mobile phone applications.

The basic blocks of RF MEMS systems are the micromechanical components like tunable capacitors, high quality inductors, microresonators, high performance filters and switches which have the potential to replace the conventional discrete components. These devices are fabricated using MEMS structures like cantilevers, air bridges, membranes and inter-digital structures.

Radio Frequency Microelectromechanical Systems

The wide application spectrum of RF MEMS devices classified as per the application domain is shown in figure. The popular RF transceiver architecture has been in existence since second world war largely due to its uncomplicated design methodology. The transceiver architecture is divided into three main blocks, the RF front-end block consisting of antenna, single-pole double- throw switch (SPDT), high performing band-pass filter, RF amplifier, voltage controlled oscillator (VCO)5, mixer, IF filter block and base band block, where analog-to-digital signal conversion takes place. While a lot of technological evolution has helped in the integration of millions of transistors without significant increase in digital block size, the RF front end still continues to rely on discrete, bulky off-chip components that interface with the digital block at PCB level. In essence, these offchip components are becoming bottleneck for system-level miniaturisation. The search for technological solutions that offer miniaturisation of off-chip components without compromising on the performance continues. After years of effort, it now appears that the RF MEMS can offer solutions to the system-level challenges presented by transceivers.

The RF MEMS provide components with reduced size and weight, very low loss, low power consumption, wide bandwidth, higher linearity, lower phase noise, better phase stability and high isolation. Wherever the application demands these features, MEMS can be conceptualised to provide solutions to replace either components or circuits or the subsystems employing the components.

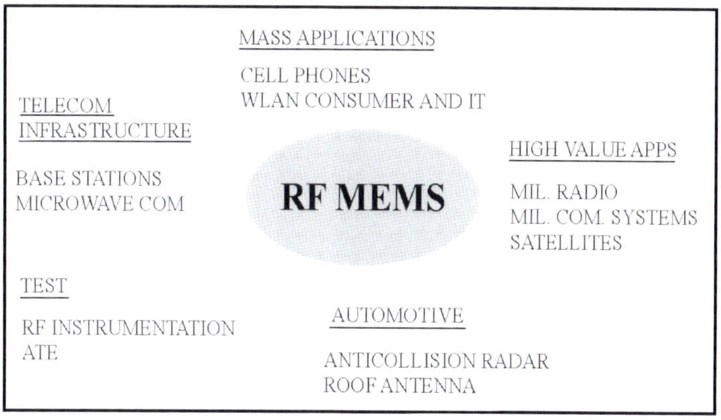

Application spectrum of RF MEMS devices.

There are a variety of RF MEMS components which are either used directly for replacement or integrated to form a microsystem along with other semiconductor devices. The components can also go along with silicon technology or *GaAs* technology and the MEMS components can be integrated to provide a system solution. The limitations faced by the normal RF integrated devices can be overcome by the flexibility and better device performance characteristics of RF MEMS components, which ultimately propagate the device level benefits to the system to attain the unprecedented levels of performance. The component level to system level development of a typical communication system using RF MEMS devices is shown in figure.

Radio Frequency Microelectromechanical Systems Components

The probable candidates which can give leverage over the conventional components in terms of high quality, low power consumption and wider operational frequency range are the tunable capacitors, high Q inductors, high performance filters, oscillators and mixers (Table 1). These components can be fabricated either by way of pre-CMOS or postCMOS approaches to integrate the MEMS components on the same wafer.

One of the key advantages offered by MEMS is the precision tunability and the tuning ratio where in the conventional components fall apart6. This provides the ability to fine tune the circuits to the best optimum level and achieve the highest performance. Re-configurability is another desirable feature used to reconfigure the same antennas for various frequencies without shifting from one antenna to another. This helps in faster scanning and better directivity of the antennas. The switches, which are characterized by very low loss, low actuation voltages and better reliability are widely used for reconfiguring the antennas, routing networks, tunable filters, etc. The RF MEMS devices have extremely high linearity, which means that these create less harmonics and this feature makes them excellent candidates for broad band communication system applications, especially needing high dynamic range of operation.

Table. Classification of radio frequency microelectromechanical systems devices as per the application domain

Devices	Wireless	WLAN	GPS	Instrumentation	RFID	Radar	Missiles
Switch	***	***	**	***	**	***	***
MEMS	**	**	***				
Inductors							
Tunable		**	**		**	***	***
Capacitors							
Resonators		***	***	***	*	***	**
T-Lines						**	
MEMTENNA					**	***	**
Phase shifter						***	***
VCOS				***			**

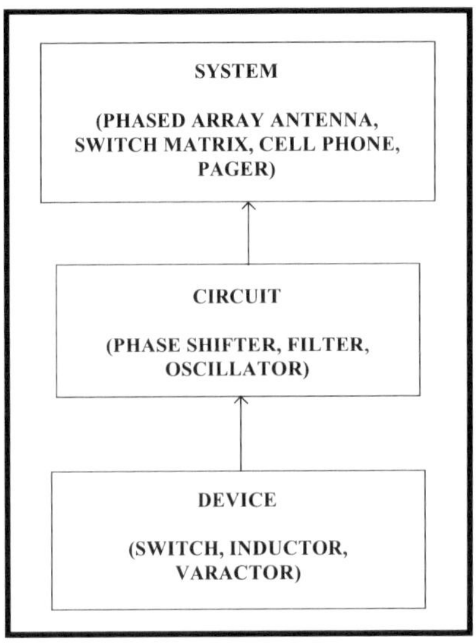

Figure: Schematic of RF MEMS system flow.

Tunable Capacitors

The standard semiconductor IC technology can provide a fixed capacitance with a sandwiched dielectric layer between two conductive electrodes In this case, the parasitic capacitance and the series resistance will result in losses and reduction in quality factor. Requirement for achieving very high capacitances also demands a very large chip area. The off chip components provide better solution where in high dielectric materials can be used to achieve better values of capacitance.

Tunability of the capacitors is required in some of the analog circuits like voltage controlled oscillators and tuning circuits which are normally met using reverse-biased semiconductor p-n diode junction. However the semiconductor varactor diodes suffer from nonlinearity and losses, thus not very suitable for applications involving stringent requirements imposed by defence and aerospace applications. The micromachined tunable capacitors meet the requirements to a great extent.

The tunable capacitors can be realised either by surface micromachining or by bulk micromachining. However the surface micromachined capacitors are simple and basically implemented by having a bottom electrode on the substrate and a suspended electrode on the top with an air gap. The tunability is achieved by the displacement of the top membrane by applying an electrostatic force between the two plates. The top membrane is suspended by means of meander beams which act like springs and provide the deflection. The tuning range, tuning voltage, quality factor, and self-resonant frequency are the factors to be considered during the design. Various configurations of micromachined tunable capacitors are shown in figure.

Tunable capacitors are made either using suspended membranes with meander9 anchors to provide better spring constant, and thereby, low actuation voltage requirement. Various configurations of meander lines like single, double beams are used depending on the other dimensions of the structures.

Alternatively, the bulk micromachined interdigitated structure or comb structures are also used to achieve good tunable capacitors. The SEM photograph of an interdigital varactor is shown in figure. In this configuration, a spring supports a set of movable fingers that mesh with a set of stationary fingers. When a DC voltage is applied, the electrostatic force attracts the movable fingers to increase the length of overlap and thus the capacitance between the fingers. The capacitance scales linearly with the number of fingers and finger thickness and is inversely proportional to the gap. Obtaining perfectly parallel fingers of the comb and release of the comb structure are the critical issues in realising the high quality device. The DRIE process is used to achieve high aspect ratio trenches which give raise to large capacitance per unit area. The process flow of DRIE is shown in figure.

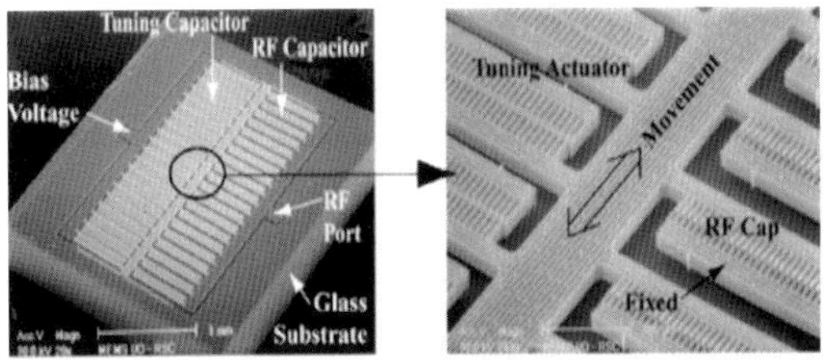

SEM picture of inter-digital varactor

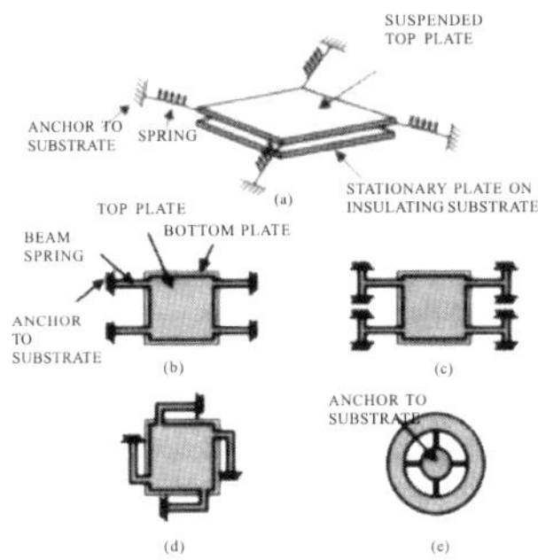

Surface micromachined tunable capacitors: (a) basic scheme, (b) top view of straight springs, (c) top view of 'T' shaped springs, (d) top view of 'L' shaped springs, and (e) top view of centre-anchored design.

The tunability is achieved by varying the bias voltage which results in the reduction of the gap between the top membrane and the bottom electrode or the gap between the two fingers in the case of IDT. The tunability ratio and the nominal capacitance are to be optimised during the design taking into account specific requirement. For very high capacitance values, a suitable dielectric with high dielectric constant can be employed in the place of air between the two plates. The parasitic capacitance can be lowered and Q can be increased by changing the low resistivity substrate like silicon to high-resistivity material like quartz.

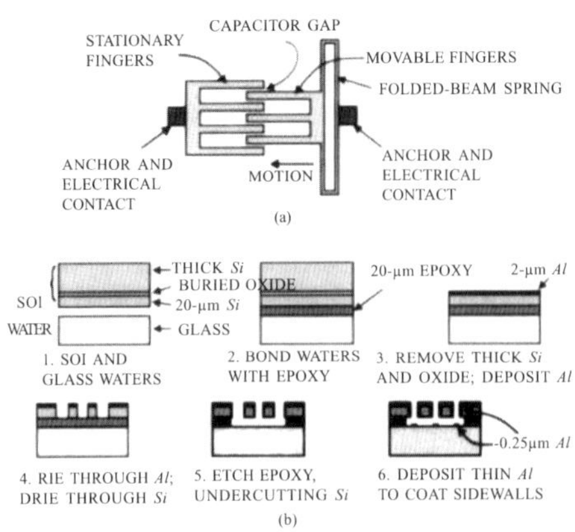

(a) Inter-digital structure (comb type) and (b) fabrication process flow

Another characteristic common to all the devices having mass and spring structures is the influence of external forces, such as acceleration and vibration which induce movement on the flexible membranes and causing corresponding change in capacitance. This result in noise and the allowed changes are application dependent.

For the sake of better damping, perforation is provided in the top electrode which also aids to remove the sacrificial layer conveniently during the process.

RF MEMS Switches

The RF MEMS switches have immense potential in consumer electronics, communication, defence and aerospace. The applications include phase shifters, switchable capacitors, tunable filters, switch matrix for routing and phased array antennas. Ideally, switches are the components whose job is to perform signal routing without causing any power loss to signal. The MEMS switches have many potential applications in electronics. The desirable parameters in RF switches are low insertion loss and return loss in the closed state, high isolation in the open state, high linearity, high power handling capability during switching, low operating voltages, high reliability, small size, and low cost. The concept of switch mechanism is illustrated in figure (a) in the open state, metal line acts as a waveguide with the sides being ground and the signal propagating down the centre line, and (b) in the closed state, application of a dc voltage pulls the top ground membrane to short the signal line. If there is a dielectric layer, the impedance is low only at high frequency.

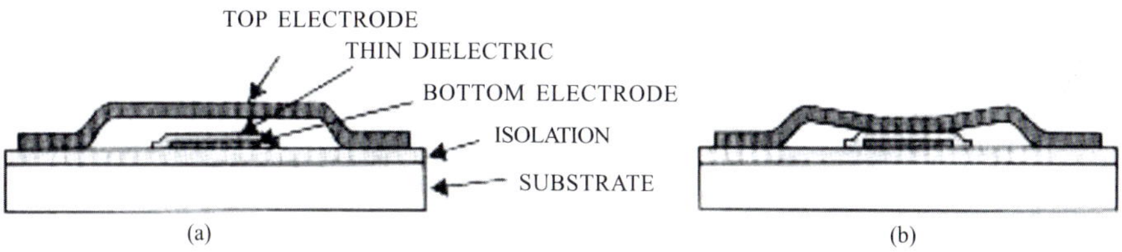

Illustration of a membrane switch in: (a) open state and (b) close state.

There are different configurations in RF MEMS switches depending on the application requirements. Normally a series ohmic switch is used while transmitting the RF signals in the frequencies dc to few GHz. For higher frequencies, the ideal solution is to use a shunt switch which provides very good isolation. Some times, the combination of these switches are conveniently used to meet wider bandwidth of operation. The key parameters considered while designing a switch are the losses, switching speeds, actuation voltages, RF power handling, and life cycles. Though various actuation mechanisms are available for actuating the MEMS membranes/ bridges/ cantilevers, the widely used mechanism in switches is the electrostatic actuation where the potential is applied between the two electrodes to bring in the displacement of the free beam. Most of these switches have a membrane or a cantilever containing one contact which is suspended over the substrate supporting another contact.

The low actuation voltages are desired, though it is a trade-off between the actuation voltage and the speed. A dielectric layer on the bottom electrode serves to cause electrostatic force to pull down the top beam. The RF MEMS switches are supposed to give better RF characteristics compared to conventional silicon *p-i-n* diodes or *GaAs* FETs and traditional relays, in terms of losses, isolation, and return loss. This makes routing of RF signals possible with much lower loss, giving RF systems better noise figure and sensitivity. The insertion loss will be affected by mismatch which comes from the beam's characteristic impedance being different from 50 Ω as well as conductor losses of the beam and dielectric losses. The isolation, on the other hand, will be affected by the beam to substrate separation that defines the 'OFF' state parasitic capacitance. The parameters, i.e., the beam to substrate separation, beam stiffness (spring constant), actuation voltage, dielectric thickness, switching speed are inter related and needs to be optimised to achieve the most desired operational parameter.

A trade-off exists between the actuation voltage and the resonant frequency for improved device performance, as low switching voltage implies low switching speeds. Also, the voltage and speed depend on switch dimensions. The voltage and resonant frequency depend inversely on the beam length. The voltage can be reduced by reducing the gap between the electrodes, but the constraint is that the off state parasitic capacitance becomes high as the gap is reduced.

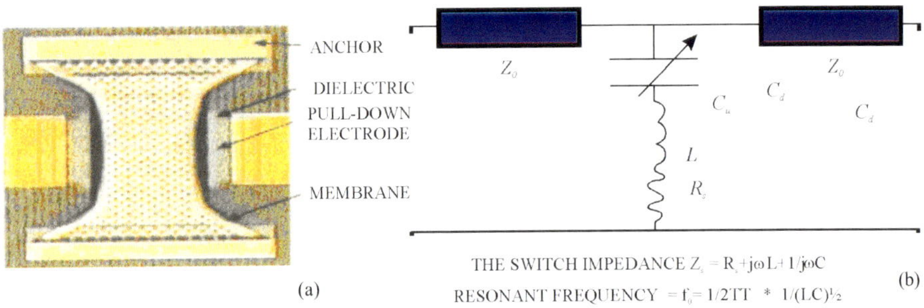

(a) Top view and (b) equivalent circuit of the RF MEMS switch.

A detailed study on RF MEMS switches has been done at Research Centre Imarat (RCI) in designing various configurations. The RF MEMS switches in ohmic configuration were designed in the frequency range DC- 7 GHz. Various configurations were tried and with different beam lengths (450-540 μm) keeping the beam width (50 μm) and thickness (1.5 μm) constant. The schematic cross-sectional view of different shunt switch configurations (single actuation pad, double actuation pad, and common actuation and transmission line) are shown in figure(a) wheras the 3-D model of the series, shunt, and combination of series and shunt switches is shown in figure.

The insertion loss, return loss, and isolation were simulated using HFSS. The modal analysis was carried in the mechanical domain using Coventorware software and the resonant frequencies were studied for different modes which falls around 12-20 KHz for different structures. Electromechanical analysis was performed and the pull-in and contact voltages were obtained. The switches were optimised to work with actuation voltages of < 15 V. The summary of the simulation results carried out at RCI is shown in tables.

MEMS Inductors

Inductors are the key elements determining the performance of the tuned circuits, in particular impedance-matching networks, low-noise amplifiers and voltage-controlled oscillators (VCOs). MEMS variable inductors find wide applications in the RF MEMS-based communication systems. The large parasitics associated with normal on chip planar inductors contributing to their lower quality factor Q can be eliminated by employing the micromachined inductors which lay above the substrate. The VCOs are potential candidates wherein MEMS inductors find place along with the tunable capacitors. In normal planar inductor, the resistive metal lines and the dielectric losses in the substrate, contribute for the degradation of the Q factor and also cause fringing and parasitic capacitances. The micromachined inductors with suspended metal structures offer very high Q factor resulting in high frequency performance of the systems.

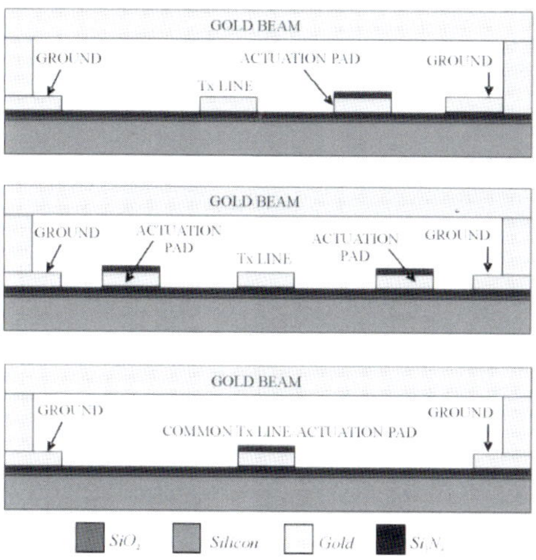

(a) Cross-sectional view of three types of shunt switch configurations

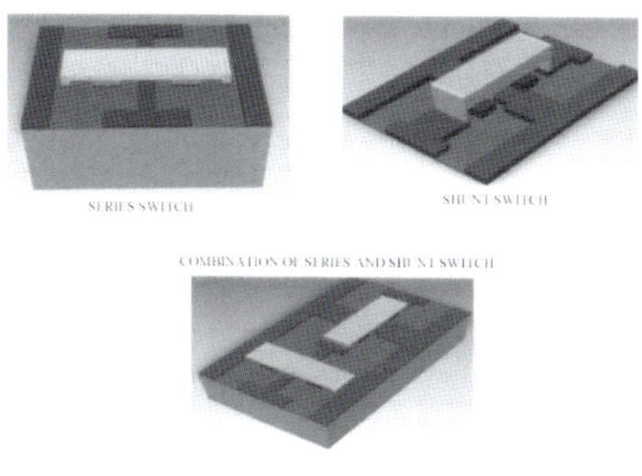

(b) 3-D pictures of switch structures.

Table: Performance comparison of switches

Switch type	Isolation	Insertion loss	Power handling	Power consumption	Switching speed
p-i-n diodes	Medium	Medium	Medium	High	Medium
GaAs FETs	Medium	Medium	Low	Medium	High
MEMS switch	High	High	Medium	Low	Low

Table: Summary of switch simulation results in electromechanical domain

Switch configuration	Suspended bridge dimensions (μm) (l x w x th)	Pull-in voltage (V)	Contact voltage (V)	c_{down} / c_{on}
	540 x 50 x 1.5	5.0	5.5	
Asymmetric				35.2
	450 x 50 x 1.5	12.0	12.5	
	540 x 50 x 1.5	5.25	5.5	
Symmetric				35.2
	450 x 50 x 1.5	8.75	9.0	
	540 x 50 x 1.5	3.0	6.25	
Common RF / actuation pad				41.25
	450 x 50 x 1.5	5.0	7.5	

Table. Summary of switch simulation results in RF domain

Switch configuration	Suspended bridge dimensions (μm) (l x w x th)	Frequency range (8 – 26 GHz) (dB)		
		Insertion	Isolation	Return loss
	540 x 50 x 1.5	-0.09 - -0.16	-31.24 - -23.85	-20.49 --19.74
Asymmetric				
	450 x 50 x 1.5	-0.09 - -0.21	-32.10 - -24.47	-19.52 - -18.01
	540 x 50 x 1.5	-0.02 - -0.17	-31.44 - -21.65	-33.93 - -27.61
Symmetric				
	450 x 50 x 1.5	-0.02 - -0.22	-32.13 - -21.65	-32.07 - -28.86
	540 x 50 x 1.5	-0.05 - -0.36	-26.17 - -20.00	-24.62 - -12.24
Common RF / actuation pad				
	450 x 50 x 1.5	-0.05 - -0.34	-27.60 - -20.60	-25.46 - -12.75

The inductors can be designed in two configurations either suspended spiral or as a solenoid as shown in figure. The inductor consists of a planar spiral made in one layer of metal and a connection to the center of the spiral in another layer of metal. In both the configurations, the dielectric losses due to the substrate as well as the parasitic capacitances and fringing due to the metal lines in the substrate can be avoided by suspending the structure over the substrate. In the case of solenoid, the magnetic field cannot be strictly confined in the solenoid due to small dimensions and there will be fringing magnetic fields in the substrate which induces eddy currents, leading to more losses The suspended spiral overcomes this as the metal spiral is suspended over the substrate at a reasonable height so that there cannot be any fringing effect giving rise to losses. Another way of reducing the effect of fringing is using a very high resistive silicon substrate (20000 Ω-cm). Micro machined inductors are used in low-noise oscillators, high-gain amplifiers, and matching networks. The nominal inductances will be around few nH and the Q factors around 20-30 are achieved using MEMS inductors.

MEMS Resonators

Quartz is very commonly used in most of the electrical resonant circuits as the integrated electronic oscillators cannot reach the quality factors that matches quartz, which is necessary for the stable operation of frequency in selective communication systems. When one takes the normal electrical RLC circuits, the quality factor is limited to < 100 because of the parasitic resistive losses in the circuit. This has also an impact on the insertion losses due to which the signal suffers an undesirable attenuation. The micromachined resonators with *Q* factors much above 1000 over a wide range of tunable frequencies when used in conjunction with the integrated electronics will result in highly stable communication system. The frequencies of interest cover the range between 800 MHz and 2.5 GHz for front-end wireless reception as well as the intermediate frequencies at 455 KHz and above.

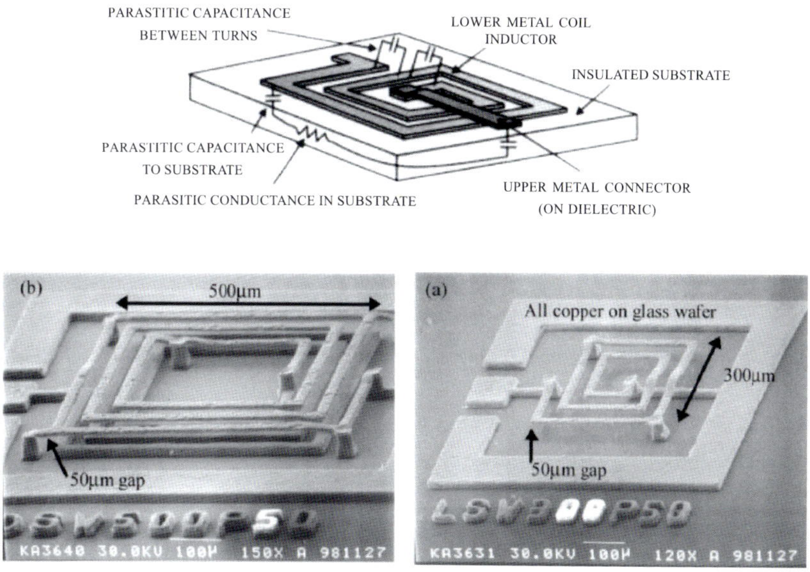

Planar on-chip inductor with parasitic noted

The bulk cavity resonators exhibit a very high *Q* exceeding 10,000. However, dimensions are very large especially in the low frequency range up to few hundreds of MHz. The micromachined resonators can be realised in dimensions that are order of magnitude smaller than cavity resonators, at the same time, meeting the high quality requirements in addition to low loss and high linearity.

There are two types of MEMS resonators, one with the vertical excitation using cantilever beams and the other with lateral excitation using comb like structures. The inter-digitated-finger comb drive structures shown in figure are commonly used in many of the devices. The applied voltage generates electrostatic force between the left anchor comb and the shuttle comb and pulls the shuttle plate to the left varying the overlap area resulting in a change in capacitance. But both these suffer from the limitation of operational frequency which can be extended only up to few hundreds of MHz only.

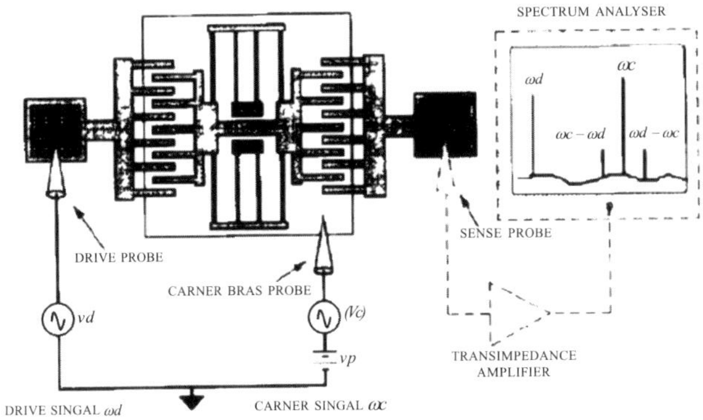

Illustration of comb resonator structure

The limitation in this comb structure is that the selfresonant frequency is restricted and related to spring constant and motional mass. The spring constant can be increased to some extent by changing the dimensions of the beam width (increase) and the beam length (decrease). The resonant frequency can also be increased using materials with high stiffness-to-mass ratio, such as aluminum, titanium or may be silicon carbide.

Another class of resonators is the beam resonator with higher self-resonant frequency achieved through bulk micromachining. This type of resonators can give higher self-resonant frequencies, better linearity in frequency variation with temperature and smaller size. A simple beam resonator is shown in figure. When a DC voltage is applied between the beam and the underlying drive electrode, the centre of the beam deflects downward and the capacitance between the flexible beam and the drive electrode varies with the deflection.

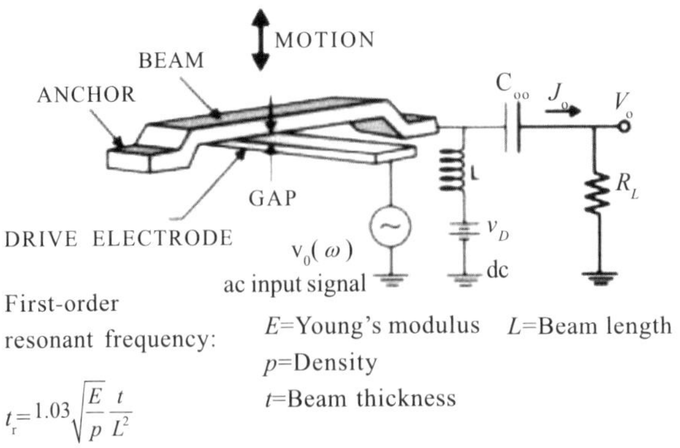

Illustration of transverse resonator structure.

Both the above configurations have the limitation to achieve higher frequencies. For applications requiring higher frequencies, that is up to few GHz, the film bulk acoustic resonator (FBAR) offers better solution. An FBAR basically consists of a thin layer of piezoelectric material, sandwiched between two electrodes. An alternating electrical potential is applied to its surfaces. Depending on the electric field and polarisation of piezoelectric layer, the thickness of the stack will change and an acoustic wave is generated in the bulk of the film. At certain frequencies, the piezoelectric layer will show resonating behaviour similar to that of the quartz material and can be used for oscillator/filter applications. This type of resonators can go upto 2 GHz with quality factors reaching over 1000.

RF MEMS Subsystems

Various subsystems have been conceived using the RF MEMS components with superior performance characteristics which cumulatively leads to the higher performance of the system. These MEMS components can be put as drop in replacement in the system or can be made an integral part of the system by fabricating the components on the same wafer. Whatever may be the approach, the systems are going to derive the benefits of these MEMS components and a number of applications are envisaged using this technology. A few applications which have a great impact on defence and aerospace are phase shifters, transreceivers, and antennas.

Phase Shifters

A phase shifter is a two port network, wherein the phase difference between output and input signal can be controlled by a control signal, usually dc bias. Phase shifters with low insertion loss, low drive power, continuous tunability and low production cost are the key to the development of light weight phased array antennas. The phased array antennas are widely used for telecommunications and Radar applications serving both the civilian and military needs. Presently the phase shifters are based on ferrite materials, p-i-n diodes or FET switches providing a good planar solution for application in microwave frequencies (However they introduce high losses which are undesirable while conceiving strategic systems).

There are two approaches to realise the MEMS phase shifters, one with switched line and the other with distributed MEMS transmission line (DMTL). The phase shifter with switched line approach has MEMS switches in the configuration which when selectively actuated results in desired phase shift while the DMTL phase shifter can be optimised to achieve low insertion loss over a wide frequency band. The phase shifters based on RF MEMS switches results in substantial reduction in losses than their solid state counter parts. This is highly favored for systems meant for onboard applications where in the power demand is a constraint. The reduction in losses reduces the demand for the MW power and thereby reduces the size apart from offering better performance. This means that one can eliminate an amplifier stage in the Tx/Rx chain. Though this may not be an issue for ground based systems, for airborne systems and space based systems this is definitely a big advantage. The well established approach is to simply replace the solid state switches with MEMS switches.

The phase shifters are designed using the delay lines and the MEMS switches to introduce the desired delay by correspondingly activating the relevant switches in the transmission line. The number of bits depend on the resolution required and no. of steps required which will be achieved by different combinations of the delay bits. For example, a 3-bit phase shifter is based on the 45/90/180 set of delay networks and can provide phase shifts of 0°, 45°, 90°, 135°, 180°, 225°, 270°, and 315° depending on the combination of bits used. The scanning resolution and side lobe levels of a phased array antenna is directly related to the number of bits employed. High performance systems may require higher number of bits like a 5-bit or 6-bit phase shifters. The insertion loss, linearity, phase accuracy and band width requirements are taken into account while designing the device. Figure shows a 5-bit phase shifter designed by authors laboratory to achieve a 360° phase shift in 32 steps. The frequency of operation is in Ku band.

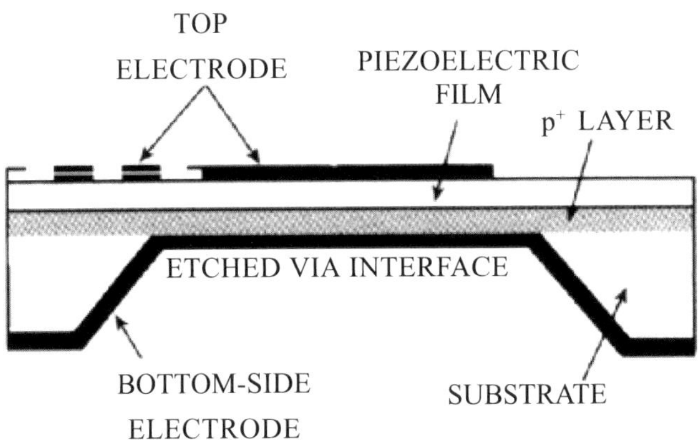

Cross-sectional view of film bulk acoustic resonator structure.

The MEMS switches are used in each segment of the delay line and are actuated separately by providing dc bias. The effect of MEMS switches in phase shifter design is that they result in lower loss phase shifters at any frequency especially from 8 GHz to 100 GHz. Also since MEMS switches have very small up-state capacitances, they result in wider band performance than similar designs using solid state devices. The switches are shunt switches designed and optimized for best RF properties, which cumulatively contributes for the overall phase shifter performance. Finally, MEMS switches can be fabricated directly with the antenna element on ceramic, quartz, or teflon-based substrates, thereby resulting in low-cost phased arrays.

The switched path phase shifters are inherently digitally switchable with only a finite number of discrete states and become cumbersome with increasing switched paths.

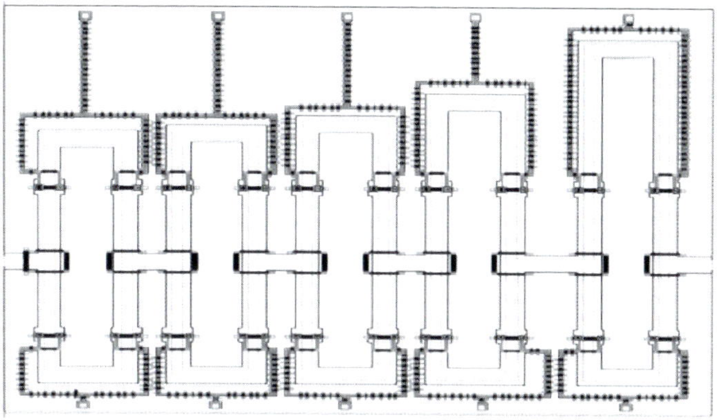

Switched line phase shifter.

Distributed MEMS transmission line (DMTL) is another approach to realise a phase shifter where in the transmission line is capacitively loaded periodically with MEMS bridges along the line and the phase velocity and phase shift can be varied accordingly. The gap between the air bridge and the bottom T line can be varied by applying electrostatic force. By controlling the bridge height, the distributed capacitance on the T line, phase velocity and phase shift can be varied and a phase shift of 0-3600 can be achieved. Here, the phase can varied continually (analog) and is more rugged than switch based.They offer very wideband performance and can work well at very high frequencies (W band and beyond). This type of phase shifter can be used to continuously vary the phase shift depending on the phase shift per unit length.

The phase shifters find application in telecommunication and military radars. Our interest in phase shifters stems from its application for the electronically scanable antennas which can replace the conventional mechanically steered antennas. They have very broad applicability for both commercial and military applications including advanced military radars, satellite communication, cellular base stations and automotive anticollision radars. Electronically scanned phased array antennas have many benefits over its mechanical counterpart in terms of faster scanning rates, reduced mechanical complexity, ability to host multiple antenna beams on the same array, wider operational frequency apart from higher reliability.

The phase shifters are highly potential devices and these are extensively used in phased array antennas in defence for a variety of applications like smart munitions, missile communication, drones, missile radar seekers, aircraft and helicopters, ground electronic steerable antennas (ESA).The requirement runs into few millions over the few years when this is fully exploited for the defence applications.

RF MEMS Transceiver

Another possible application of RF MEMS is their implementation in transceivers in wireless systems. The demand for portable wireless systems with more functionality is facilitating rapid growth in the technology to come out with newer devices with greater functional density and with integratability into the system. RF MEMS have made their impact with their potential in transceivers associated with very low power consumption, high quality factor, and size reduction. Infact, applications like personal communication systems (PCS), wireless networking, radars are being fully exploited by MEMS to reduce the parts count, power consumption, and lighter weight coupled with superior RF performance. Several types of RF MEMS devices such as switches, variable capacitors (varactors), inductors, and filters, have been developed and the technology is maturing to the level of deployment in commercial systems. Though these RF MEMS devices can be developed as discrete components, using many RF devices will increase the size of RF transceivers. Another problem is related to packaging. These RF devices containing movable parts have to be individually packaged which will result in the increase of overall size. The solution to exploit fully the benefit of RF MEMS is to go for the integrated RF MEMS on a single wafer and wafer level encapsulation process can be used for the packaging in an optimum size. The configuration for integrated RF MEMS technology is conceptually shown in figure.

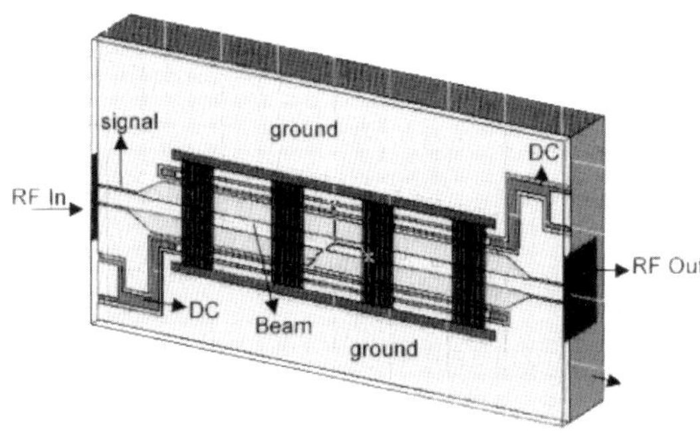

3-D view of designed DMTL phase shifter at RCI

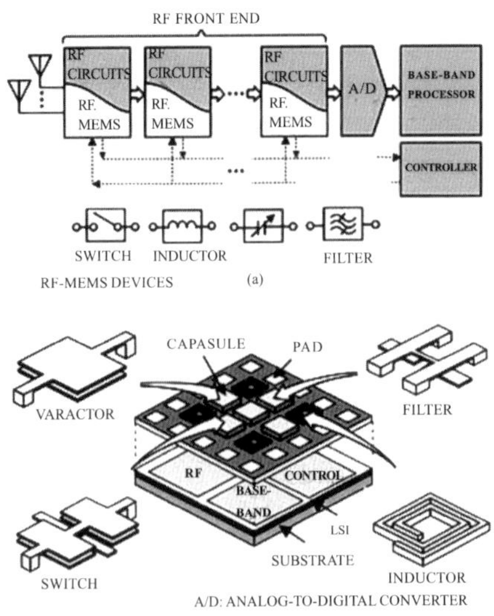

Conceptual design of integrated RF MEMS frontend receiver

This configuration reduces the volume consumed by discrete packaging of the devices and allows the use of a lot of devices without increasing the numbcr of off-chip components. RF MEMS devices such as switches with movable parts are protected by a device scale capsule to prevent their destruction during packaging. This technology will lead to the fusion of RF-MEMS and circuit technologies and to the development of single-chip reconfigurable RF transceivers.

The general block diagram of transceiver for RF communication indicating the possible application of RF MEMS components is shown in figure below. This will result in systems with better RF performance, multi-band communicability, smaller size and with lower power consumption. The focus on this technology is primarily because of its large potential in future mobile communication enabling the device to carry out multifunctions.

RF MEMS Packaging

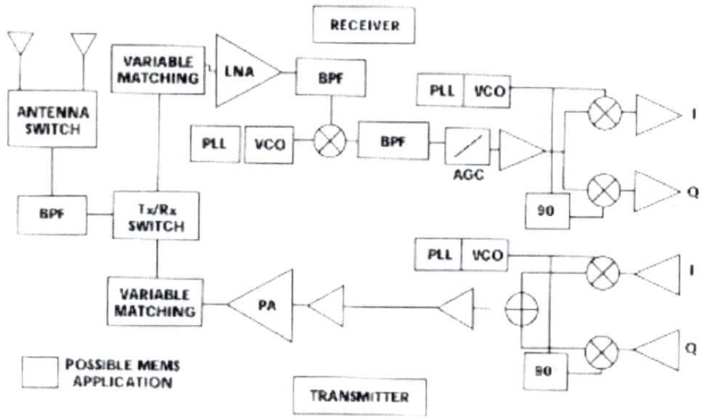

Conventional receiver architecture indicating the possible MEMS application

Packaging of MEMS is another critical bottleneck area in the realisation of MEMS product. Unlike the case of ICs, MEMS are sensitive to environmental conditions. The free-standing structures are fragile and must be properly encapsulated. An effective package protects the movable structure as well as prevent particles and contaminants from hindering their operation. The purpose of the package is to protect the device from all aspects of environment while interfering with the device performance in minimal way. For RF devices, it means the package needs to be invisible in the desired signal path (low insertion loss), be opaque to RF in every other possible signal path (good isolation, low cross-talk), defend the device from internal and external heat problems (including solderability, temperature cycling, power handling, etc.), defend it from environmental problems (fatigue, shock, vibration) and defend the device from internal and external contamination (hermiticity, gettering, etc.). In some cases, it also calls for vacuum sealing as the air causes viscous damping leading to reduction of Q factor in RF MEMS components. The package aspects are designed to ensure RF performance and at the same time provide the levels of operational characteristics like life time, repeatability, reliability, and ruggedness which are points of high concern for military and space applications.

In addition to ensuring of unwanted resonances and electromagnetic interference and coupling, RF MEMS packaging techniques aim at preventing moisture and particulates, which may impair the movement of freestanding MEMS structures, leading to change in resonant frequencies and more losses.

Surface Acoustic Wave Sensor

Surface acoustic wave (SAW) resonators represent some of the most prominent acoustic devices for chemical sensing applications. As their frequency ranges from several hundred MHz to GHz, therefore they can record remarkably diminutive frequency shifts resulting from exceptionally small mass loadings. Their miniaturized design, high thermal stability and possibility of wireless integration make these devices highly competitive. Owing to these special characteristics, they are widely accepted as smart transducers that can be combined with a variety of recognition layers based on host-guest interactions, metal oxide coatings, carbon nanotubes, graphene sheets, functional polymers and biological receptors. As a result of this, there is a broad spectrum of SAW sensors, i.e., having sensing applications ranging from small gas molecules to large bio-analytes or even whole cell structures.

Device Layout

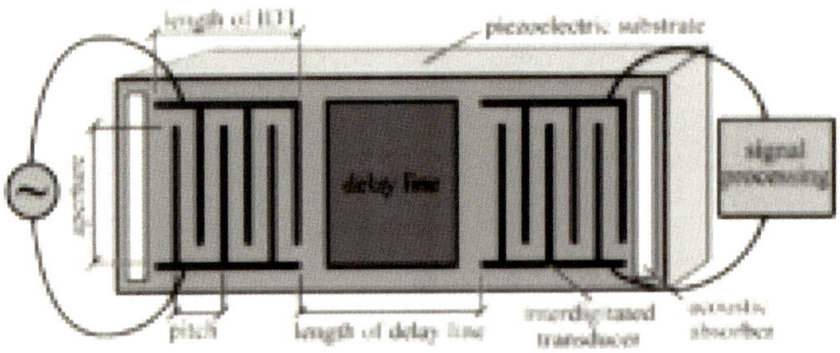

Surface Acoustic Wave Sensor Interdigitated Transducer Diagram

The basic surface acoustic wave device consists of a piezoelectric substrate, an input interdigitated transducer (IDT) on one side of the surface of the substrate, and a second, output interdigitated transducer on the other side of the substrate. The space between the IDTs, across which the surface acoustic wave will propagate, is known as the delay-line. This region is called the delay line because the signal, which is a mechanical wave at this point, moves much slower than its electromagnetic form, thus causing an appreciable delay.

Device Operation

Surface acoustic wave technology takes advantage of the piezoelectric effect in its operation. Most modern surface acoustic wave sensors use an input interdigitated transducer (IDT) to convert an electrical signal into an acoustic wave.

The sinusoidal electrical input signal creates alternating polarity between the fingers of the interdigitated transducer. Between two adjacent sets of fingers, polarity of the fingers will be switched (e.g. + - +). As a result, the direction of the electric field between two fingers will alternate between adjacent sets of fingers. This creates alternating regions of tensile and compressive strain between fingers of the electrode by the piezoelectric effect, producing a mechanical wave at the surface known as a surface acoustic wave. As fingers on the same side of the device will be at the same level of compression or tension, the space between them-known as the pitch-is the wavelength of the mechanical wave. We can express the synchronous frequency f_0 of the device with phase velocity υ_p and pitch p as:

—

The synchronous frequency is the natural frequency at which mechanical waves should propagate. Ideally, the input electric signal should be at the synchronous frequency to minimize insertion loss.

As the mechanical wave will propagate in both directions from the input IDT, half of the energy of the waveform will propagate across the delay line in the direction of the output IDT. In some devices, a mechanical absorber or reflector is added between the IDTs and the edges of the substrate to prevent interference patterns or reduce insertion losses, respectively.

The acoustic wave travels across the surface of the device substrate to the other interdigitated transducer, converting the wave back into an electric signal by the piezoelectric effect. Any changes that were made to the mechanical wave will be reflected in the output electric signal. As the characteristics of the surface acoustic wave can be modified by changes in the surface properties of the device substrate, sensors can be designed to quantify any phenomenon which alters these properties. Typically, this is accomplished by the addition of mass to the surface or changing the length of the substrate and the spacing between the fingers.

Inherent Functionality

The structure of the basic surface acoustic wave sensor allows for the phenomena of pressure, strain, torque, temperature, and mass to be sensed. The mechanisms for this are discussed below:

Pressure, Strain, Torque, Temperature

The phenomena of pressure, strain, torque, temperature, and mass can be sensed by the basic device, consisting of two IDTs separated by some distance on the surface of a piezoelectric substrate. These phenomena can all cause a change in length along the surface of the device. A change in length will affect both the spacing between the interdigitated electrodes-altering the pitch-and the spacing between IDTs-altering the delay. This can be sensed as a phase-shift, frequency-shift, or time-delay in the output electrical signal.

When a diaphragm is placed between the environment at a variable pressure and a reference cavity at a fixed pressure, the diaphragm will bend in response to a pressure differential. As the diaphragm bends, the distance along the surface in compression will increase. A surface acoustic wave pressure sensor simply replaces the diaphragm with a piezoelectric substrate patterned with interdigitated electrodes. Strain and torque work in a similar manner, as application to the sensor will cause a deformation of the piezoelectric substrate. A surface acoustic wave temperature sensor can be fashioned from a piezoelectric substrate with a relatively high coefficient of thermal expansion in the direction of the length of the device.

Mass

The accumulation of mass on the surface of an acoustic wave sensor will affect the surface acoustic wave as it travels across the delay line. The velocity v of a wave traveling through a solid is proportional to the square root of product of the Young's modulus E and the density of the material.

$$v \propto \sqrt{E / \rho}$$

Therefore, the wave velocity will decrease with added mass. This change can be measured by a change in time-delay or phase-shift between input and output signals. Signal attenuation could be measured as well, as the coupling with the additional surface mass will reduce the wave energy. In the case of mass-sensing, as the change in the signal will always be due to an increase in mass from a reference signal of zero additional mass, signal attenuation can be effectively used.

Extended Functionality

The inherent functionality of a surface acoustic wave sensor can be extended by the deposition of a thin film of material across the delay line which is sensitive to the physical phenomena of interest. If a physical phenomenon causes a change in length or mass in the deposited thin film, the surface acoustic wave will be affected by the mechanisms mentioned above. Some extended functionality examples are listed below:

Chemical Vapors

Chemical vapor sensors use the application of a thin film polymer across the delay line which selectively absorbs the gas or gases of interest. An array of such sensors with different polymeric coatings can be used to sense a large range of gases on a single sensor with resolution down to parts per trillion, allowing for the creation of a sensitive "lab on a chip."

Biological Matter

A biologically-active layer can be placed between the interdigitated electrodes which contains immobilized antibodies. If the corresponding antigen is present in a sample, the antigen will bind to the antibodies, causing a mass-loading on the device. These sensors can be used to detect bacteria and viruses in samples, as well as to quantify the presence of certain mRNA and proteins.

Humidity

Surface acoustic wave humidity sensors require a thermoelectric cooler in addition to a surface acoustic wave device. The thermoelectric cooler is placed below the surface acoustic wave device. Both are housed in a cavity with an inlet and outlet for gases. By cooling the device, water vapor will tend to condense on the surface of the device, causing a mass-loading.

Ultraviolet Radiation

Surface acoustic wave devices can be made sensitive to optical wavelengths through the phenomena known as acoustic charge transport (ACT), which involves the interaction between a surface acoustic wave and photogenerated charge carriers from a photoconducting layer. Ultraviolet radiation sensors employ the use of a thin film layer of zinc oxide across the delay line. When exposed to ultraviolet radiation, zinc oxide generates charge carriers which interact with the electric fields produced in the piezoelectric substrate by the traveling surface acoustic wave. This interaction decreases the velocity and the amplitude of the signal.

Magnetic Fields

Ferromagnetic materials, such as iron, nickel, and cobalt, exhibit a characteristic called magnetostriction, where the Young's modulus of the material is dependent on magnetic field strength. If a constant stress is maintained on such a material, the strain will change with a changing Young's modulus. If such a material is deposited in the delay line of a surface acoustic wave sensor, a change in length of the deposited film will stress the underlying substrate. This stress will result in a strain on the surface of the substrate, affecting the phase velocity, phase-shift, and time-delay of the signal.

Viscosity

Surface acoustic wave devices can be used to measure changes in viscosity of a liquid placed upon it. As the liquid becomes more viscous the resonant frequency of the device will change in correspondence. A network analyser is needed to view the resonant frequency.

Smart Dust

Smart Dust devices are small wireless microelectromechanical sensors (MEMS) that can detect everything from light to vibrations. It is a tiny dust size device with extraordinary capabilities. It encompasses nano-structured silicon sensor which can spontaneously assemble, orient sense and report on their local environment. This new technology combines sensing, computing, wireless communication capabilities and autonomous power supply within the volume of only a few millimeters. It is very hard to detect the presence of the Smart Dust and it is even harder to get rid of them once deployed. Smart Dust is useful in monitoring real world phenomenon without disturbing the original process.

Key Components of Smart Dust

- A semiconductor laser diode and MEMS beam steering mirror for active optical transmission
- Corner Cube Retro reflector (CCR) for passive optical transmission
- Photodetector and receiver
- An optical receiver
- A signal processing and control circuitry
- A power source based on thick-film batteries and solar cells.

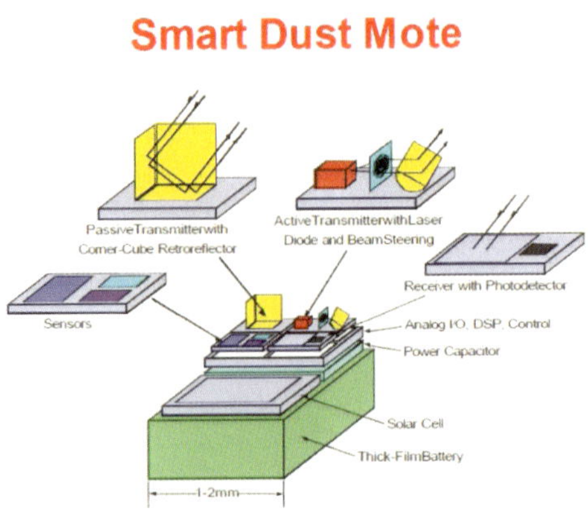

Construction of Smart Dust

The new trend of the 21st century is to provide the best adaptive features to the user in small packages as much as possible. This same trend is being followed in electronics technology and the enhance features are being provided in smaller and smaller electronic devices. An example of this technology trend is electronic motes. These are so small and light in weight that they can remain suspended in the environment like an ordinary dust particle. Even the air currents can also move them in the direction of flow. The main features of motes are:

- Support the collection and integration of data from a variety of miniature sensors.
- Analyze the sensor data as specified by system level controls. Wirelessly communicate the results of their analyses to other motes, system base stations, and the internet as specified by system automation.
- Motes are also sometimes referred to as smart dust. One mote is composed of a small, low powered and cheap computer connected to several sensors and a radio transmitter capable of forming ad hoc networks. The computer monitors the different sensors in a mote. These sensors can measure light, acceleration, position, stress, pressure, humidity, sound, and vibration. Data gathered are passed on to the radio link for transmission from mote to mote until data reaches the transmission node.

Working Principle of Smart Dust

Smart Dust motes are run by microcontrollers. These microcontrollers consist of tiny sensors for recording various type of data. Timers are used to run these sensors. These sensors do the job of collecting the data. The data obtained are stored in its memory for further interpretations. It can also be sent to the base controlling stations.

CCR, that comprises of three mutually perpendicular mirrors of gold coated polysilicon, has the property that any incident ray of light is reflected back to the source provided that is incident within a certain range of angles centered about the cubes body diagonal. The microfabricated CCR includes an electrostatic actuator that can deflect one of the mirrors at kilohertz rate.Hence, the external light source can be transmitted back in the form of the modulated signal at kilobits per second. It can transmit to the bus only when the CCR body diagonal happens to point directly towards the bits, within a few tens of degrees.

Although a passive transmitter can be made more omnidirectional by employing several CCR's oriented in different directions, at the expense of increased dust mote size.

Advantages of Smart Dust

There are many advantages to using Smart Dust.

For an industry:

- Improving safety, efficiency, and compliance
- Reducing system and infrastructure costs
- Increasing productivity

For farmers or farming purpose:

- Detecting the needs of the crop resulting in a better fertilization management.
- It gives farmers a better management of time.
- Reduction of inputs and Increase of outputs resulting an increase in productivity.

For factories and plants:

- It automates many manual error-prone tasks which involve calibration and monitoring.
- Provide accurate data of motor health in order to perform more timely maintenance when needed
- Detection of corrosion in aging pipes before they leak

For an office environment:

- It eliminates wired routers entirely and replacing them with a single Smart Dust chip which would handle all hardware and software functions for distributed networks, using five times less power than conventional networks
- Smart Dust nodes can even be equipped with GPS receivers. Can be used to track the movements of visitors as they roam around the office to see if they are going to any restricted locations.

For military purpose:

- A military application like monitoring activities in inaccessible areas, accompany soldiers and alert them to any poisons or dangerous biological substances in the air.
- Security and Tracking
- In the military, they can perform as a remote sensor chip to track enemy movements, detect poisonous gas or radioactivity.

References

- Waldner, Jean-Baptiste (2008). Nanocomputers and Swarm Intelligence. London: ISTEJohn Wiley & Sons. p. 205. ISBN 1-84821-009-4.
- Microelectromechanical-system: britannica.com, Retrieved 27 May 2018
- Wilfinger, R.J.; Bardell, P.H.; Chhabra, D.S. (1968). "The Resonistor: A Frequency Selective Device Utilizing the Mechanical Resonance of a Silicon Substrate". IBM J. 12: 113–8. doi:10.1147/rd.121.0113.
- Introduction-microelectromechanical-systems-mems: compliantmechanisms.byu.edu, Retrieved 17 July 2018
- By Peter Clarke, EE Times Europe. "Smart MEMS microphones market emerges." May 31, 2016. Retrieved June 1, 2016.
- Nanoelectromechanical Systems: aldebaran.cz, Retrieved 14 April 2018
- Williams, K.R.; Muller, R.S. (1996). "Etch rates for micromachining processing" (PDF). Journal of Microelectromechanical Systems. 5 (4): 256. doi:10.1109/84.546406.
- R. Ghodssi; P. Lin (2011). MEMS Materials and Processes Handbook. Berlin: Springer. ISBN 978-0-387-47316-1.
- Nanotechnology, technology: britannica.com, Retrieved 11 June 2018
- Kovacs, G.T.A.; Maluf, N.I.; Petersen, K.E. (1998). "Bulk micromachining of silicon"(PDF). Proceedings of the IEEE. 86 (8): 1536. doi:10.1109/5.704259.

2

Microfabrication

Microfabrication is the technology of fabricating structures at the micrometer or even smaller scales. The processes of doping, etching, thin films, microlithography, photolithography, etc. are integral to the development of microfabrication technology. The topics included in this chapter on surface micromachining, photolithography, etching, thin-film deposition, wafer dicing, etc. are intended to provide an easy understanding of the varied aspects of microfabrication.

Surface Micromachining

Surface micromachining is the fabrication of micromechanical structures by deposition and etching of thin structural and sacrificial films. Thus, simple microstructures like beams or membranes as well as complex structures like linkages or encapsulated resonators can be fabricated on top of a silicon substrate. Surface micromachining is characterized by the fabrication of micromechanical structures from deposited thin films. Originally employed for integrated circuits, films composed of materials such as low-pressure chemical-vapor-deposition polycrystalline silicon, silicon nitride, and silicon dioxides can be sequentially deposited and selectively removed to build or "machine" three-dimensional structures whose functionality typically requires that they be freed from the planar substrate. Although the process to accomplish this fabrication dates from the 1960's, its rapid extension over the past few years and its application to batch fabrication of micromechanisms and of monolithic microelectromechanical systems (MEMS) make a thorough review of surface micromachining appropriate at this time. Four central issues of consequence to the MEMS technologist are: (i) the understanding and control of the material properties of microstructural films, such as polycrystalline silicon, (ii) the release of the microstructure, for example, by wet etching silicon dioxide sacrificial films, followed by its drying and surface passivation, (iii) the constraints defined by the combination of micromachining and integrated-circuit technologies when fabricating monolithic sensor devices, and (iv) the methods, materials, and practices used when packaging the completed device. Last, recent developments of hinged structures for postrelease assembly, high-aspect-ratio fabrication of molded parts from deposited thin films, and the advent of deep anisotropic silicon etching hold promise to extend markedly the capabilities of surface-micromachining technologies.

Microfabrication

Microfabrication is the process of fabrication of miniature structures of micrometre scales and smaller. Historically, the earliest microfabrication processes were used for integrated circuit fabrication, also known as "semiconductor manufacturing" or "semiconductor device fabrication". In the last two decades microelectromechanical systems, microsystems, micromachines and their subfields, microfluidics/lab-on-a-chip, optical MEMS, RF MEMS, Power MEMS, Bio MEMS and their extension into nanoscale have re-used, adapted or extended microfabrication methods. Flat-panel displays and solar cells are also using similar techniques. Miniaturization of various devices presents challenges in many areas of science and engineering: physics, chemistry, materials science, computer science, ultra-precision engineering, fabrication processes, and equipment design. It is also giving rise to various kinds of interdisciplinary research. The major concepts and principles of microfabrication are microlithography, doping, thin films, etching, bonding, and polishing.

Fields of use

Microfabricated devices include:

- Fabrication of integrated circuits ("microchips")
- Microelectromechanical systems (MEMS), MOEMS,
- microfluidic devices (ink jet print heads)
- solar cells
- Flat Panel Displays
- Sensors (micro-sensors) (biosensors, nanosensors)
- PowerMEMSs, fuel cells, energy harvesters/scavengers

Microfabrication technologies originate from the microelectronics industry, and the devices are usually made on silicon wafers even though glass, plastics and many other substrate are in use. Micromachining, semiconductor processing, microelectronic fabrication, semiconductor fabrication, MEMS fabrication and integrated circuit technology are terms used instead of microfabrication, but microfabrication is the broad general term.

Traditional machining techniques such as *electro-discharge machining*, *spark erosion machining*, and *laser drilling* have been scaled from the millimeter size range to micrometer range, but they do not share the main idea of microelectronics-originated microfabrication: replication and parallel fabrication of hundreds or millions of identical structures. This parallelism is present in various imprint, casting and moulding techniques which have successfully been applied in the microregime. For example, injection moulding of DVDs involves fabrication of submicrometer-sized spots on the disc.

Processes

Microfabrication is actually a collection of technologies which are utilized in making microdevices. Some of them have very old origins, not connected to manufacturing, like lithography or etching. Polishing was borrowed from optics manufacturing, and many of the vacuum techniques come from 19th century physics research. Electroplating is also a 19th-century technique adapted to produce micrometre scale structures, as are various stamping and embossing techniques.

To fabricate a microdevice, many processes must be performed, one after the other, many times repeatedly. These processes typically include depositing a film, patterning the film with the desired micro features, and removing (or etching) portions of the film. Thin film metrology is used typically during each of these individual process steps, to ensure the film structure has the desired characteristics in terms of thickness (t), refractive index (n) and extinction coefficient (k), for suitable device behavior. For example, in memory chip fabrication there are some 30 lithography steps, 10 oxidation steps, 20 etching steps, 10 doping steps, and many others are performed. The complexity of microfabrication processes can be described by their *mask count*. This is the number of different pattern layers that constitute the final device. Modern microprocessors are made with 30 masks while a few masks suffice for a microfluidic device or a laser diode. Microfabrication resembles multiple exposure photography, with many patterns aligned to each other to create the final structure.

Substrates

Microfabricated devices are not generally freestanding devices but are usually formed over or in a thicker support substrate. For electronic applications, semiconducting substrates such as silicon wafers can be used. For optical devices or flat panel displays, transparent substrates such as glass or quartz are common. The substrate enables easy handling of the micro device through the many fabrication steps. Often many individual devices are made together on one substrate and then singulated into separated devices toward the end of fabrication.

Deposition or Growth

Microfabricated devices are typically constructed using one or more thin films. The purpose of these thin films depends on the type of device. Electronic devices may have thin films which are conductors (metals), insulators (dielectrics) or semiconductors. Optical devices may have films which are reflective, transparent, light guiding or scattering. Films may also have a chemical or mechanical purpose as well as for MEMS applications. Examples of deposition techniques include:

- Thermal oxidation
- Local oxidation of Silicon (LOCOS)
- Chemical vapor deposition (CVD)
  - APCVD
  - LPCVD
  - PECVD
- Physical vapor deposition(PVD)
  - Sputtering
  - Evaporative deposition
  - Electron beam PVD
- Epitaxy

Patterning

It is often desirable to pattern a film into distinct features or to form openings (or vias) in some of the layers. These features are on the micrometer or nanometer scale and the patterning technology is what defines microfabrication. The patterning technique typically uses a 'mask' to define portions of the film which will be removed. Examples of patterning techniques include:

- Photolithography
- Shadow Masking

Etching

Etching is the removal of some portion of the thin film or substrate. The substrate is exposed to an etching (such as an acid or plasma) which chemically or physically attacks the film until it is removed. Etching techniques include:

- Dry etching (Plasma etching) such as Reactive-ion etching (RIE) or Deep reactive-ion etching (DRIE)
- Wet etching or Chemical Etching

Microforming

Microforming is a microfabrication process of microsystem or microelectromechanical system (MEMS) "parts or structures with at least two dimensions in the submillimeter range." It includes techniques such as microextrusion, microstamping, and microcutting. These and other microforming processes have been envisioned and researched since at least 1990, leading to the development of industrial- and experimental-grade manufacturing tools. However, as Fu and Chan pointed out, several issues must still be resolved before the technology can be implemented more widely, including deformation load and defects, forming system stability, mechanical properties, and other size-related effects on the crystallite (grain) structure and boundaries.

In microforming, the ratio of the total surface area of grain boundaries to the material volume decreases with the decrease of specimen size and the increase of grain size. This leads to the decrease of grain boundary strengthening effect. Surface grains have lesser constraints compared to internal grains. The change of flow stress with part geometry size is partly attributed to the change of volume fraction of surface grains. In addition, the anisotropic properties of each grain become significant with the decrease of workpiece size, which results in the inhomogeneous deformation, irregular formed geometry and the variation of deformation load. There is a critical need to establish the systematic knowledge of microforming to support the design of part, process, and tooling with the consideration of size effects.

Other

A wide variety of other processes for cleaning, planarizing, or modifying the chemical properties of microfabricated devices can also be performed. Some examples include:

- Doping by either thermal diffusion or ion implantation
- Chemical-mechanical planarization (CMP)
- Wafer cleaning, also known as "surface preparation"
- Wire bonding

Cleanliness in Wafer Fabrication

Microfabrication is carried out in cleanrooms, where air has been filtered of particle contamination and temperature, humidity, vibrations and electrical disturbances are under stringent control. Smoke, dust, bacteria and cells are micrometers in size, and their presence will destroy the functionality of a microfabricated device.

Cleanrooms provide passive cleanliness but the wafers are also actively cleaned before every critical step. RCA-1 clean in ammonia-peroxide solution removes organic contamination and particles; RCA-2 cleaning in hydrogen chloride-peroxide mixture removes metallic impurities. Sulfuric acid-peroxide mixture (a.k.a. Piranha) removes organics. Hydrogen fluoride removes native oxide from silicon surface. These are all wet cleaning steps in solutions. Dry cleaning methods include oxygen and argon plasma treatments to remove unwanted surface layers, or hydrogen bake at elevated temperature to remove native oxide before epitaxy. Pre-gate cleaning is the most critical cleaning step in CMOS fabrication: it ensures that the ca. 2 nm thick oxide of a MOS transistor can be grown in an orderly fashion. Oxidation, and all high temperature steps are very sensitive to contamination, and cleaning steps must precede high temperature steps.

Surface preparation is just a different viewpoint, all the steps are the same: it is about leaving the wafer surface in a controlled and well known state before you start processing. Wafers are contaminated by previous process steps (e.g. metals bombarded from chamber walls by energetic ions during ion implantation), or they may have gathered polymers from wafer boxes, and this might be different depending on wait time.

Wafer cleaning and surface preparation work a little bit like the machines in a bowling alley: first they remove all unwanted bits and pieces, and then they reconstruct the desired pattern so that the game can go on.

Photolithography

The word lithography comes from the Greek *lithos*, meaning stones, and *graphia*, meaning to write. It means quite literally writing on stones. In the case of semiconductor lithography (also called photolithography) our stones are silicon wafers and our patterns are written with a light sensitive polymer called a photoresist. To build the complex structures that make up a transistor and the many wires that connect the millions of transistors of a circuit, lithography and etch pattern transfer steps are repeated at least 10 times, but more typically are done 20 to 30 times to make one circuit. Each pattern being printed on the wafer is aligned to the previously formed patterns and slowly the conductors, insulators, and selectively doped regions are built up to form the final device.

The importance of lithography can be appreciated in two ways. First, due to the large number of lithography steps needed in IC manufacturing, lithography typically accounts for about 30 percent of the cost of manufacturing. Second, lithography tends to be the technical limiter for further advances in feature size reduction and thus transistor speed and silicon area. Obviously, one must carefully understand the trade-offs between cost and capability when developing a lithography process. Although lithography is certainly not the only technically important and challenging process in the IC manufacturing flow, historically, advances in lithography have gated advances in IC cost and performance.

Optical lithography is basically a photographic process by which a light sensitive polymer, called a photoresist, is exposed and developed to form three-dimensional relief images on the substrate. In general, the ideal photoresist image has the exact shape of the designed or intended pattern in the plane of the substrate, with vertical walls through the thickness of the resist. Thus, the final resist pattern is binary: parts of the substrate are covered with resist while other parts are completely uncovered. This binary pattern is needed for pattern transfer since the parts of the substrate covered with resist will be protected from etching, ion implantation, or other pattern transfer mechanism.

The general sequence of processing steps for a typical photolithography process is as follows: substrate preparation, photoresist spin coat, prebake, exposure, post-exposure bake, development, and postbake. A resist strip is the final operation in the lithographic process, after the resist pattern has been transferred into the underlying layer. This sequence is shown diagrammatically in figure, and is generally performed on several tools linked together into a contiguous unit called a *lithographic cluster*.

Basic Procedure

A single iteration of photolithography combines several steps in sequence. Modern cleanrooms use automated, robotic wafer track systems to coordinate the process. The procedure described here omits some advanced treatments, such as thinning agents or edge-bead removal.

Cleaning

If organic or inorganic contaminations are present on the wafer surface, they are usually removed by wet chemical treatment, e.g. the RCA clean procedure based on solutions containing hydrogen peroxide. Other solutions made with trichloroethylene, acetone or methanol can also be used to clean.

Preparation

The wafer is initially heated to a temperature sufficient to drive off any moisture that may be present on the wafer surface, 150°C for ten minutes is sufficient. Wafers that have been in storage must be chemically cleaned to remove contamination. A liquid or gaseous "adhesion promoter", such as Bis(trimethylsilyl)amine ("hexamethyldisilazane", HMDS), is applied to promote adhesion of the photoresist to the wafer. The surface layer of silicon dioxide on the wafer reacts with HMDS to form tri-methylated silicon-dioxide, a highly water repellent layer not unlike the layer of wax on a car's paint. This water repellent layer prevents the aqueous developer from penetrating between the photoresist layer and the wafer's surface, thus preventing so-called lifting of small photoresist structures in the (developing) pattern. In order to ensure the development of the image, it is best covered and placed over a hot plate and let it dry while stabilizing the temperature at 120°C.

Photoresist Application

The wafer is covered with photoresist by spin coating. A viscous, liquid solution of photoresist is dispensed onto the wafer, and the wafer is spun rapidly to produce a uniformly thick layer. The spin coating typically runs at 1200 to 4800 rpm for 30 to 60 seconds, and produces a layer between 0.5 and 2.5 micrometres thick. The spin coating process results in a uniform thin layer, usually with uniformity of within 5 to 10 nanometres. This uniformity can be explained by detailed fluid-mechanical modelling, which shows that the resist moves much faster at the top of the layer than at the bottom, where viscous forces bind the resist to the wafer surface. Thus, the top layer of resist is quickly ejected from the wafer's edge while the bottom layer still creeps slowly radially along the wafer. In this way, any 'bump' or 'ridge' of resist is removed, leaving a very flat layer. Final thickness is also determined by the evaporation of liquid solvents from the resist. For very small, dense features (< 125 or so nm), lower resist thicknesses (< 0.5 micrometres) are needed to overcome collapse effects at high aspect ratios; typical aspect ratios are < 4:1.

The photo resist-coated wafer is then prebaked to drive off excess photoresist solvent, typically at 90 to 100°C for 30 to 60 seconds on a hotplate.

Exposure and Developing

After prebaking, the photoresist is exposed to a pattern of intense light. The exposure to light causes a chemical change that allows some of the photoresist to be removed by a special solution, called "developer" by analogy with photographic developer. Positive photoresist, the most common type, becomes soluble in the developer when exposed; with negative photoresist, unexposed regions are soluble in the developer.

A post-exposure bake (PEB) is performed before developing, typically to help reduce standing wave phenomena caused by the destructive and constructive interference patterns of the incident light. In deep ultraviolet lithography, chemically amplified resist (CAR) chemistry is used. This process is much more sensitive to PEB time, temperature, and delay, as most of the "exposure" reaction (creating acid, making the polymer soluble in the basic developer) actually occurs in the PEB.

The develop chemistry is delivered on a spinner, much like photoresist. Developers originally often contained sodium hydroxide (NaOH). However, sodium is considered an extremely undesirable contaminant in MOSFET fabrication because it degrades the insulating properties of gate oxides (specifically, sodium ions can migrate in and out of the gate, changing the threshold voltage of the transistor and making it harder or easier to turn the transistor on over time). Metal-ion-free developers such as tetramethylammonium hydroxide (TMAH) are now used.

The resulting wafer is then "hard-baked" if a non-chemically amplified resist was used, typically at 120 to 180°C for 20 to 30 minutes. The hard bake solidifies the remaining photoresist, to make a more durable protecting layer in future ion implantation, wet chemical etching, or plasma etching.

Etching

In etching, a liquid ("wet") or plasma ("dry") chemical agent removes the uppermost layer of the substrate in the areas that are not protected by photoresist. In semiconductor fabrication, dry etching techniques are generally used, as they can be made anisotropic, in order to avoid significant undercutting of the photoresist pattern. This is essential when the width of the features to be defined is similar to or less than the thickness of the material being etched (i.e. when the aspect ratio approaches unity). Wet etch processes are generally isotropic in nature, which is often indispensable for microelectromechanical systems, where suspended structures must be "released" from the underlying layer.

The development of low-defectivity anisotropic dry-etch process has enabled the ever-smaller features defined photolithographically in the resist to be transferred to the substrate material.

Photoresist Removal

After a photoresist is no longer needed, it must be removed from the substrate. This usually requires a liquid "resist stripper", which chemically alters the resist so that it no longer adheres to the substrate. Alternatively, photoresist may be removed by a plasma containing oxygen, which oxidizes it. This process is called ashing, and resembles dry etching. Use of 1-Methyl-2-pyrrolidone (NMP) solvent for photoresist is another method used to remove an image. When the resist has been dissolved, the solvent can be removed by heating to 80°C without leaving any residue.

Exposure ("Printing") Systems

Wafer track portion of an aligner that uses 365 nm ultraviolet light

Exposure systems typically produce an image on the wafer using a photomask. The photomask blocks light in some areas and lets it pass in others. (Maskless lithography projects a precise beam directly onto the wafer without using a mask, but it is not widely used in commercial processes.) Exposure systems may be classified by the optics that transfer the image from the mask to the wafer.

Photolithography produces better thin film transistor structures than printed electronics, due to smoother printed layers, less wavy patterns, and more accurate drain-source electrode registration.

Contact and Proximity

A contact printer, the simplest exposure system, puts a photomask in direct contact with the wafer and exposes it to a uniform light. A proximity printer puts a small gap between the photomask and wafer. In both cases, the mask covers the entire wafer, and simultaneously patterns every die.

Contact printing is liable to damage both the mask and the wafer, and this was the primary reason it was abandoned for high volume production. Both contact and proximity lithography require the light intensity to be uniform across an entire wafer, and the mask to align precisely to features already on the wafer. As modern processes use increasingly large wafers, these conditions become increasingly difficult.

Research and prototyping processes often use contact or proximity lithography, because it uses inexpensive hardware and can achieve high optical resolution. The resolution in proximity lithography is approximately the square root of the product of the wavelength and the gap distance. Hence, except for projection lithography, contact printing offers the best resolution, because its gap distance is approximately zero (neglecting the thickness of the photoresist itself). In addition, nanoimprint lithography may revive interest in this familiar technique, especially since the cost of ownership is expected to be low; however, the shortcomings of contact printing discussed above remain as challenges.

Projection

Very-large-scale integration (VLSI) lithography uses projection systems. Unlike contact or proximity masks, which cover an entire wafer, projection masks (known as "reticles") show only one die or an array of dies (known as a "field"). Projection exposure systems (steppers or scanners) project the mask onto the wafer many times to create the complete pattern.

Photomasks

The image for the mask originates from a computerized data file. This data file is converted to a series of polygons and written onto a square of fused quartz substrate covered with a layer of chromium using a photolithographic process. A laser beam (laser writer) or a beam of electrons (e-beam writer) is used to expose the pattern defined by the data file and travels over the surface of the substrate in either a vector or raster scan manner. Where the photoresist on the mask is exposed, the chrome can be etched away, leaving a clear path for the illumination light in the stepper/scanner system to travel through.

Resolution in Projection Systems

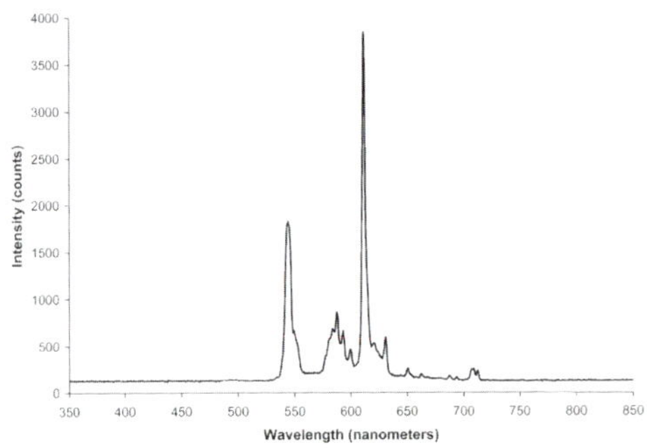

The filtered fluorescent lighting in photolithography cleanrooms contains no ultraviolet or blue light in order to avoid exposing photoresists. The spectrum of light emitted by such fixtures gives virtually all such spaces a bright yellow color.

The ability to project a clear image of a small feature onto the wafer is limited by the wavelength of the light that is used, and the ability of the reduction lens system to capture enough diffraction orders from the illuminated mask. Current state-of-the-art photolithography tools use deep ultraviolet (DUV) light from excimer lasers with wavelengths of 248 and 193 nm (the dominant lithography technology today is thus also called "excimer laser lithography"), which allow minimum feature sizes down to 50 nm. Excimer laser lithography has thus played a critical role in the continued advance of the so-called Moore's Law for the last 20 years.

The minimum feature size that a projection system can print is given approximately by:

$$CD = k_1 \cdot \frac{\lambda}{NA}$$

where

CD is the minimum feature size (also called the critical dimension, *target design rule*). It is also common to write 2 *times* the *half-pitch*.

k_1 (commonly called *k1 factor*) is a coefficient that encapsulates process-related factors, and typically equals 0.4 for production. The minimum feature size can be reduced by decreasing this coefficient through Computational lithography.

λ is the wavelength of light used

NA is the numerical aperture of the lens as seen from the wafer

According to this equation, minimum feature sizes can be decreased by decreasing the wavelength, and increasing the numerical aperture (to achieve a tighter focused beam and a smaller spot size). However, this design method runs into a competing constraint. In modern systems, the depth of focus is also a concern:

$$D_F = k_2 \cdot \frac{\lambda}{NA^2}$$

Here, k_2 is another process-related coefficient. The depth of focus restricts the thickness of the photoresist and the depth of the topography on the wafer. Chemical mechanical polishing is often used to flatten topography before high-resolution lithographic steps.

Light Sources

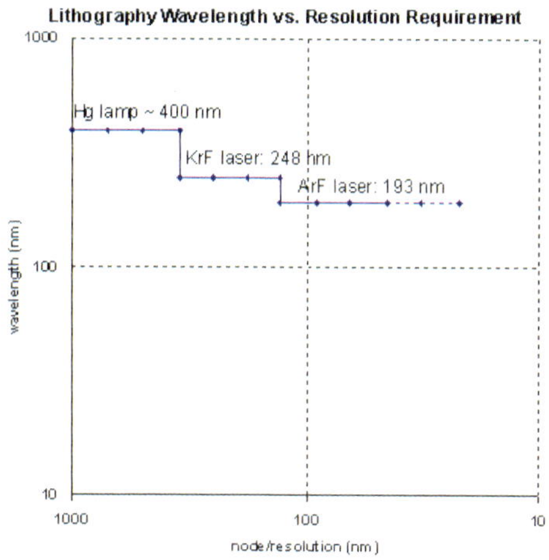

One of the evolutionary paths of lithography has been the use of shorter wavelengths. It is worth noting that the same light source may be used for several technology generations.

Historically, photolithography has used ultraviolet light from gas-discharge lamps using mercury, sometimes in combination with noble gases such as xenon. These lamps produce light across a broad spectrum with several strong peaks in the ultraviolet range. This spectrum is filtered to select a single spectral line. From the early 1960s through the mid-1980s, Hg lamps had been used in lithography for their spectral lines at 436 nm ("g-line"), 405 nm ("h-line") and 365 nm ("i-line"). However, with the semiconductor industry's need for both higher resolution (to produce denser and faster chips) and higher throughput (for lower costs), the lamp-based lithography tools were no longer able to meet the industry's high-end requirements.

This challenge was overcome when in a pioneering development in 1982, excimer laser lithography was proposed and demonstrated at I.B.M. by Kanti Jain, and now excimer laser lithography machines (steppers and scanners) are the primary tools used worldwide in microelectronics production. With phenomenal advances made in tool technology in the last two decades, it is the semiconductor industry view that excimer laser lithography has been a crucial factor in the continued advance of Moore's Law, enabling minimum features sizes in chip manufacturing to shrink from 0.5 micrometer in 1990 to 45 nanometers and below in 2010. This trend has continued into this decade for even denser chips, with minimum features reaching 10 nanometers in 2016. From an even broader scientific and technological perspective, in the 50-year history of the laser since its first demonstration in 1960, the invention and development of excimer laser lithography has been recognized as a major milestone.

The commonly used deep ultraviolet excimer lasers in lithography systems are the krypton fluoride laser at 248 nm wavelength and the argon fluoride laser at 193 nm wavelength. The primary manufacturers of excimer laser light sources in the 1980s were Lambda Physik (now part of Coherent, Inc.) and Lumonics. Since the mid-1990s Cymer Inc. has become the dominant supplier of excimer laser sources to the lithography equipment manufacturers, with Gigaphoton Inc. as their closest rival. Generally, an excimer laser is designed to operate with a specific gas mixture; therefore, changing wavelength is not a trivial matter, as the method of generating the new wavelength is completely different, and the absorption characteristics of materials change. For example, air begins to absorb significantly around the 193 nm wavelength; moving to sub-193 nm wavelengths would require installing vacuum pump and purge equipment on the lithography tools (a significant challenge). Furthermore, insulating materials such as silicon dioxide, when exposed to photons with energy greater than the band gap, release free electrons and holes which subsequently cause adverse charging.

Optical lithography has been extended to feature sizes below 50 nm using the 193 nm ArF excimer laser and liquid immersion techniques. Also termed immersion lithography, this enables the use of optics with numerical apertures exceeding 1.0. The liquid used is typically ultra-pure, deionised water, which provides for a refractive index above that of the usual air gap between the lens and the wafer surface. The water is continually circulated to eliminate thermally-induced distortions. Water will only allow *NA*'s of up to ~1.4, but fluids with higher refractive indices would allow the effective *NA* to be increased further.

Experimental tools using the 157 nm wavelength from the F2 excimer laser in a manner similar to current exposure systems have been built. These were once targeted to succeed 193 nm lithography at the 65 nm feature size node but have now all but been eliminated by the introduction of immersion lithography. This was due to persistent technical problems with the 157 nm technology and economic considerations that provided strong incentives for the continued use of 193 nm excimer laser lithography technology. High-index immersion lithography is the newest extension of 193 nm lithography to be considered. In 2006, features less than 30 nm were demonstrated by IBM using this technique.

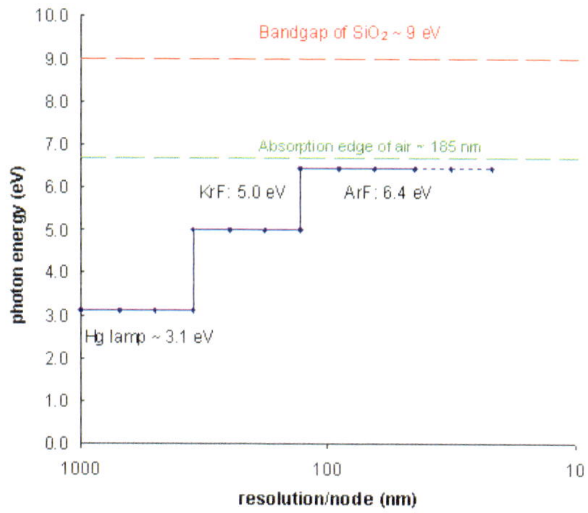

Changing the lithography wavelength is significantly limited by absorption. Air absorbs below c. 185 nm.

UV excimer lasers have been demonstrated to about 126 nm (for Ar_2^*). Mercury arc lamps are designed to maintain a steady DC current of 50 to 150 Volts, however excimer lasers have a higher resolution. Excimer lasers are gas-based light systems that are usually filled with inert and halide gases (Kr, Ar, Xe, F and Cl) that are charged by an electric field. The higher the frequency, the greater the resolution of the image. KrF lasers are able to function at a frequency of 4 kHz. In addition to running at a higher frequency, excimer lasers are compatible with more advanced machines than mercury arc lamps are. They are also able to operate from greater distances (up to 25 meters) and are able to maintain their accuracy with a series of mirrors and antireflective-coated lenses. By setting up multiple lasers and mirrors, the amount of energy loss is minimized, also since the lenses are coated with antireflective material, the light intensity remains relatively the same from when it left the laser to when it hits the wafer.

Lasers have been used to indirectly generate non-coherent extreme UV (EUV) light at 13.5 nm for extreme ultraviolet lithography. The EUV light is not emitted by the laser, but rather by a tin or xenon plasma which is excited by an excimer laser. Fabrication of feature sizes of 10 nm has been demonstrated in production environments, but not yet at rates needed for commercialization. However, this is expected by 2016. This technique does not require a synchrotron, and EUV sources, as noted, do not produce coherent light. However, vacuum systems and a number of novel technologies (including much higher EUV energies than are now produced) are needed to work with UV at the edge of the X-ray spectrum (which begins at 10 nm).

Theoretically, an alternative lightsource for photolithography, especially if and when wavelengths continue to decrease to extreme UV or X-ray, is the free-electron laser (or one might say xaser for an X-ray device). Free-electron lasers can produce high quality beams at arbitrary wavelengths.

Visible and infrared femtosecond lasers were also applied for lithography. In that case photochemical reactions are initiated by multiphoton absorption. Usage of these light sources have a lot of benefits, including possibility to manufacture true 3D objects and process non-photosensitized (pure) glass-like materials with superb optical resiliency.

Experimental Methods

Photolithography has been defeating predictions of its demise for many years. For instance, by the early 1980s, many in the semiconductor industry had come to believe that features smaller than 1 micrometer could not be printed optically. Modern techniques using excimer laser lithography already print features with dimensions a fraction of the wavelength of light used – an amazing optical feat. New tricks such as immersion lithography, dual-tone resist and multiple patterning continue to improve the resolution of 193 nm lithography. Meanwhile, current research is exploring alternatives to conventional UV, such as electron beam lithography, X-ray lithography, extreme ultraviolet lithography and ion projection lithography.

Thin-film Deposition

Thin-film deposition is the technology of applying a very thin film of material – between a few nanometers to about 100 micrometers, or the thickness of a few atoms – onto a "substrate" surface to be coated, or onto a previously deposited coating to form layers. Thin-film deposition manufacturing processes are at the heart of today's semiconductor industry, solar panels, CDs, disk drives, and optical devices industries.

Thin-film deposition is usually divided into two broad categories – Chemical deposition and physical vapor deposition coating systems.

Chemical deposition is when a volatile fluid precursor produces a chemical change on a surface leaving a chemically deposited coating. One example is chemical vapor deposition or CVD used to produce the highest-purity, highest-performance solid materials in the semiconductor industry today.

Physical vapor deposition refers to a wide range of technologies where a material is released from a source and deposited on a substrate using mechanical, electromechanical or thermodynamic processes. The two most common techniques of physical vapor deposition or PVD are thermal evaporation and sputtering.

Thermal Evaporation

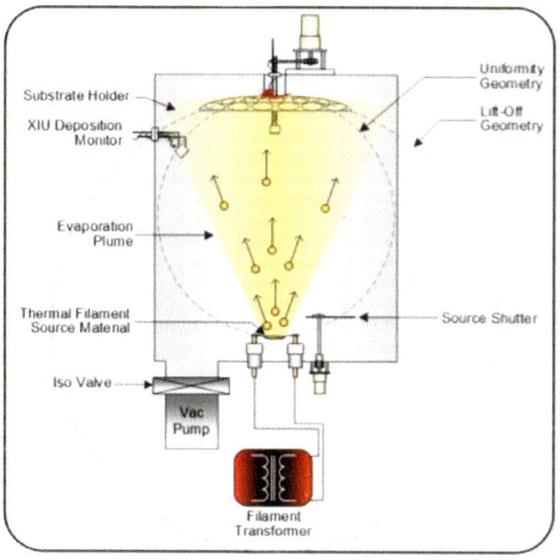

Thermal Evaporation Process

Thermal evaporation involves heating a solid material that will be used to coat a substrate inside a high vacuum chamber until it starts to boil and evaporates producing vapor pressure. Inside the vacuum deposition chamber, even a relatively low vapor pressure is sufficient to raise a vapor cloud. This evaporated material now constitutes a vapor stream which the vacuum allows to travel without reacting or scattering against other atoms. It traverses the chamber and hits the substrate, sticking to it as a coating or thin film.

There are two primary methods of heating the source material during thermal evaporation. One is known as filament evaporation, as it is achieved with a simple electrical heating element or filament. The other common heat source is an electron beam or e-beam evaporation where an electron beam is aimed at the source material to evaporate it and enter the gas phase.

Thin-film evaporation systems can offer the advantages of relatively high deposition rates, real time rate and thickness control, and (with suitable physical configuration) good evaporant stream directional control for processes such as lift off to achieve direct patterned coatings.

Sputtering

Sputtering involves the bombardment of a target material with high energy particles that are to be deposited on a substrate like a silicon wafer or solar panel. The substrates to be coated are placed in a vacuum chamber containing an inert gas – usually Argon – and a negative electric charge is placed on the target material to be deposited causing the plasma in the chamber to glow.

Atoms are "Sputtered off" the target by collisions with the Argon gas atoms, carrying these particles across the vacuum chamber and are deposited as a thin film. Several different methods of plasma vapor deposition coating systems are widely used, including ion beam and ion-assisted sputtering, reactive sputtering in an oxygen gas environment, gas flow and magnetron sputtering.

Magnetron Sputtering

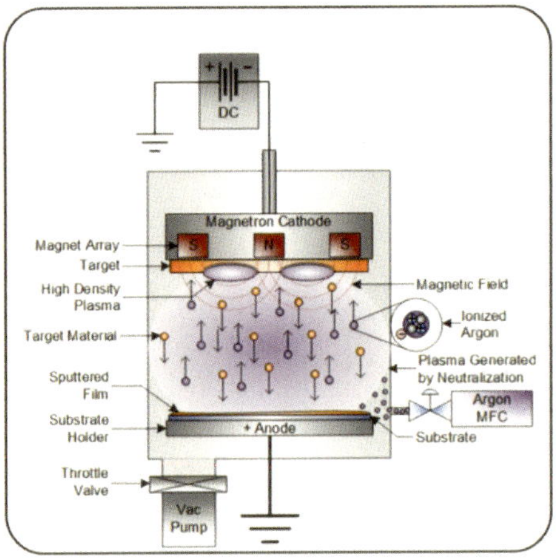

DC Magnetron Sputtering Process

Magnetron sputtering uses magnets to trap electrons over the negatively charged target material so they are not free to bombard the substrate, preventing the object to be coated from overheating or being damaged, and allowing for a faster thin film deposition rate. Magnetron Sputtering systems are typically configured as "In-line" where the substrates travel by the target material on some type of conveyor belt, or circular for smaller applications. They use several methods of inducing the high energy state including direct current (DC), alternating current (AC) and radio frequency (RF) magnetron sources.

Compared to thermal evaporation that utilizes more conventional heating temperatures, Sputtering takes place in the plasma "Fourth state of nature" environment with much higher temperatures and kinetic energies allowing a much purer and more precise thin film deposition on the atomic level.

Etching

Etching is a common practice in manufacturing and industrial processes. It is used for the chemical removal of layers from the surface of a substrate of semiconductor material, during the fabrication of ICs and wafer-based solar cells. For the manufacture of semiconductor devices and printed circuit boards, and also for the preparation of metallic specimens for microscopic observation, it is a commonly used process. Each wafer undergoes several stages of etching before its completion. For these stages, part of the wafer is masked to resist etching using a photoresist patterned with photolithography, or a durable mask like silicon nitride. If the objective of etching is to create a cavity in the material, its depth can be controlled by monitoring the etch rate or etching time. Generally, the top layer is removed without causing damage to the masking or underlying layers.

If the etch is intended to make a cavity in a material, the depth of the cavity may be controlled approximately using the etching time and the known etch rate. More often, though, etching must entirely remove the top layer of a multilayer structure, without damaging the underlying or masking layers. The etching system's ability to do this depends on the ratio of etch rates in the two materials (*selectivity*).

Some etches undercut the masking layer and form cavities with sloping sidewalls. The distance of undercutting is called *bias*. Etchants with large bias are called *isotropic*, because they erode the substrate equally in all directions. Modern processes greatly prefer anisotropic etches, because they produce sharp, well-controlled features.

Selectivity	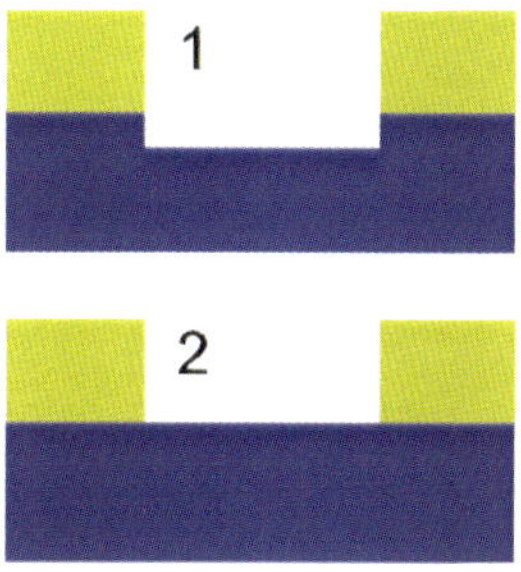	Blue: layer to remain 1. A poorly selective etch removes the top layer, but also attacks the underlying material. 2. A highly selective etch leaves the underlying material unharmed.
Isotropy	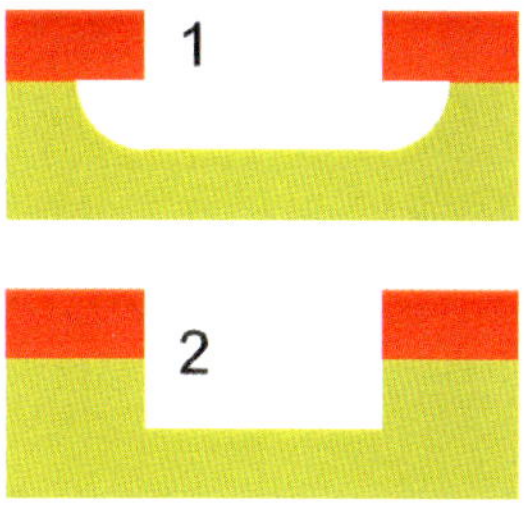	Red: masking layer; yellow: layer to be removed 1. A perfectly isotropic etch produces round sidewalls. 2. A perfectly anisotropic etch produces vertical sidewalls.

Etching Media and Technology

The two fundamental types of etchants are liquid-phase ("wet") and plasma-phase ("dry"). Each of these exists in several varieties.

Wet Etching

The first etching processes used liquid-phase ("wet") etchants. The wafer can be immersed in a bath of etchant, which must be agitated to achieve good process control. For instance, buffered hydrofluoric acid (BHF) is used commonly to etch silicon dioxide over a silicon substrate.

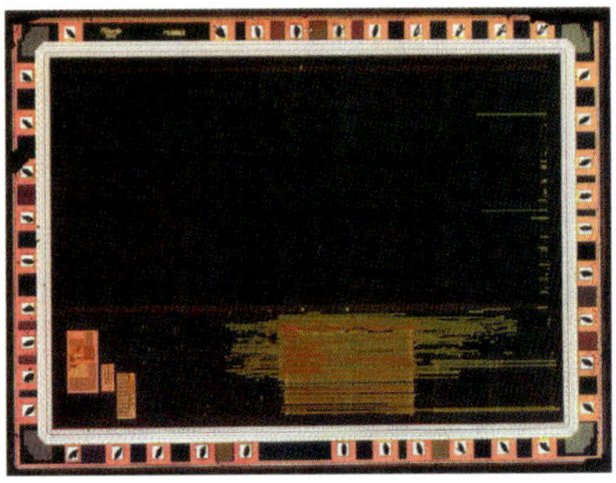

Radiation hardened die of the 1886VE10 microcontroller prior to metalization etching

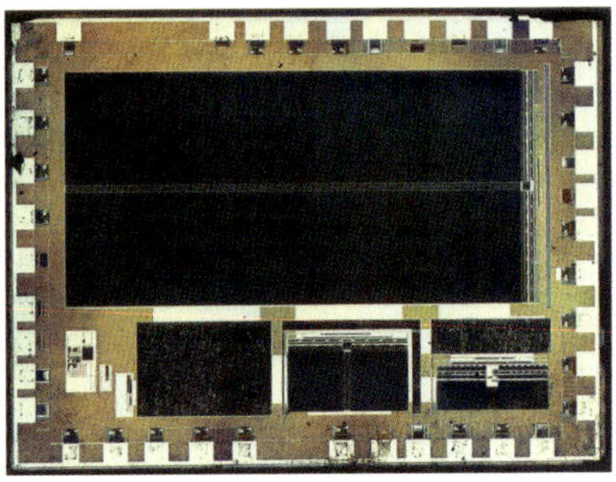

Radiation hardened die of the 1886VE10 microcontroller after a metalization etching process has been used

Different specialised etchants can be used to characterise the surface etched.

Wet etchants are usually isotropic, which leads to large bias when etching thick films. They also require the disposal of large amounts of toxic waste. For these reasons, they are seldom used in state-of-the-art processes. However, the photographic developer used for photoresist resembles wet etching.

As an alternative to immersion, single wafer machines use the Bernoulli principle to employ a gas (usually, pure nitrogen) to cushion and protect one side of the wafer while etchant is applied to the other side. It can be done to either the front side or back side. The etch chemistry is dispensed on the top side when in the machine and the bottom side is not affected. This etch method is particularly effective just before "backend" processing (BEOL), where wafers are normally very much thinner after wafer backgrinding, and very sensitive to thermal or mechanical stress. Etching a thin layer of even a few micrometres will remove microcracks produced during backgrinding resulting in the wafer having dramatically increased strength and flexibility without breaking.

Anisotropic Wet Etching (Orientation Dependent Etching)

Some wet etchants etch crystalline materials at very different rates depending upon which crystal face is exposed. In single-crystal materials (e.g. silicon wafers), this effect can allow very high anisotropy, as shown in the figure. The term "crystallographic etching" is synonymous with "anisotropic etching along crystal planes".

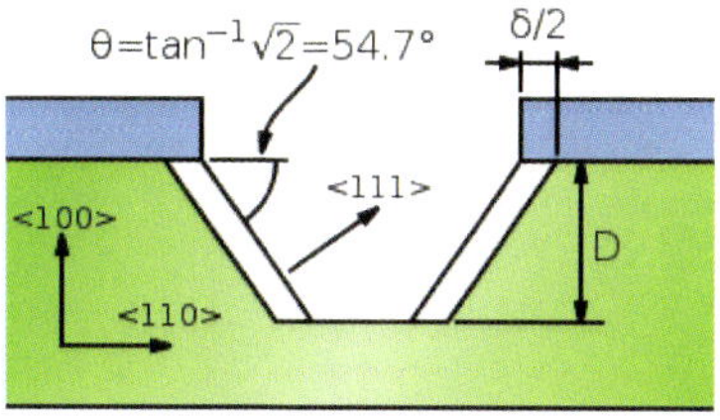

An anisotropic wet etch on a silicon wafer creates a cavity with a trapezoidal cross-section. The bottom of the cavity is a {100} plane, and the sides are {111} planes. The blue material is an etch mask, and the green material is silicon.

However, for some non-crystal materials like glass, there are unconventional ways to etch in an anisotropic manner. The authors employs multistream laminar flow that contains etching non-etching solutions to fabricate a glass groove. The etching solution at the center is flanked by non-etching solutions and the area contacting etching solutions is limited by the surrounding non-etching solutions. Thereby, the direction of etching is mainly vertical to the surface of glass. The SEM images demonstrate the breaking of conventional theoretical limit of aspect ratio (width/height=0.5) and contribute a two-fold improvement (width/height=1).

Several anisotropic wet etchants are available for silicon, all of them hot aqueous caustics. For instance, potassium hydroxide (KOH) displays an etch rate selectivity 400 times higher in <100> crystal directions than in <111> directions. EDP (an aqueous solution of ethylene diamine and pyrocatechol), displays a <100>/<111> selectivity of 17X, does not etch silicon dioxide as KOH does, and also displays high selectivity between lightly doped and heavily boron-doped (p-type) silicon. Use of these etchants on wafers that already contain CMOS integrated circuits requires protecting the circuitry. KOH may introduce mobile potassium ions into silicon dioxide, and EDP is highly corrosive and carcinogenic, so care is required in their use. Tetramethylammonium hydroxide (TMAH) presents a safer alternative than EDP, with a 37X selectivity between {100} and {111} planes in silicon.

Etching a (100) silicon surface through a rectangular hole in a masking material, for example a hole in a layer of silicon nitride, creates a pit with flat sloping {111}-oriented sidewalls and a flat (100)-oriented bottom. The {111}-oriented sidewalls have an angle to the surface of the wafer of:

$$\arctan\sqrt{2} = 54.7^{\circ}$$

If the etching is continued “to completion”, i.e. until the flat bottom disappears, the pit becomes a trench with a V-shaped cross section. If the original rectangle was a perfect square, the pit when etched to completion displays a pyramidal shape.

The undercut, δ, under an edge of the masking material is given by:

$$\delta = \frac{\sqrt{6}D}{S} = \frac{\sqrt{6}R_{100}T}{R_{100}/R_{111}} = \sqrt{6}TR_{111},$$

where R_{xxx} is the etch rate in the <xxx> direction, T is the etch time, D is the etch depth and S is the anisotropy of the material and etchant.

Different etchants have different anisotropies. Below is a table of common anisotropic etchants for silicon:

Etchant	Operating temp (°C)	R_{100} (μm/min)	$S=R_{100}/R_{111}$	Mask materials
Ethylenediamine pyrocatechol (EDP)	110	0.47	17	SiO_2, Si_3N_4, Au, Cr, Ag, Cu
Potassium hydroxide/Isopropyl alcohol (KOH/IPA)	50	1.0	400	Si_3N_4, SiO_2 (etches at 2.8 nm/min)
Tetramethylammonium hydroxide (TMAH)	80	0.6	37	Si_3N_4, SiO_2

Plasma Etching

Modern VLSI processes avoid wet etching, and use *plasma etching* instead. Plasma etchers can operate in several modes by adjusting the parameters of the plasma. Ordinary plasma etching operates between 0.1 and 5 Torr. (This unit of pressure, commonly used in vacuum engineering, equals approximately 133.3 pascals.) The plasma produces energetic free radicals, neutrally charged, that react at the surface of the wafer. Since neutral particles attack the wafer from all angles, this process is isotropic.

Plasma etching can be isotropic, i.e., exhibiting a lateral undercut rate on a patterned surface approximately the same as its downward etch rate, or can be anisotropic, i.e., exhibiting a smaller lateral undercut rate than its downward etch rate. Such anisotropy is maximized in deep reactive ion etching. The use of the term anisotropy for plasma etching should not be conflated with the use of the same term when referring to orientation-dependent etching.

The source gas for the plasma usually contains small molecules rich in chlorine or fluorine. For instance, carbon tetrachloride (CCl_4) etches silicon and aluminium, and trifluoromethane etches silicon dioxide and silicon nitride. A plasma containing oxygen is used to oxidize ("ash") photoresist and facilitate its removal.

Ion milling, or *sputter etching*, uses lower pressures, often as low as 10^{-4} Torr (10 mPa). It bombards the wafer with energetic ions of noble gases, often Ar^+, which knock atoms from the substrate by transferring momentum. Because the etching is performed by ions, which approach the wafer approximately from one direction, this process is highly anisotropic. On the other hand, it tends to display poor selectivity. *Reactive-ion etching* (RIE) operates under conditions intermediate between sputter and plasma etching (between 10^{-3} and 10^{-1} Torr). *Deep reactive-ion etching* (DRIE) modifies the RIE technique to produce deep, narrow features.

Common Etch Processes used in Microfabrication

Etchants for common microfabrication materials		
Material to be etched	**Wet etchants**	**Plasma etchants**
Aluminium (Al)	80% phosphoric acid (H_3PO_4) + 5% acetic acid + 5% nitric acid (HNO_3) + 10% water (H_2O) at 35–45°C	Cl_2, CCl_4, $SiCl_4$, BCl_3
Indium tin oxide [ITO] (In_2O_3:SnO_2)	Hydrochloric acid (HCl) + nitric acid (HNO_3) + water (H_2O) (1:0.1:1) at 40°C	
Chromium (Cr)	• "Chrome etch": ceric ammonium nitrate ($(NH_4)_2Ce(NO_3)_6$) + nitric acid (HNO_3) • Hydrochloric acid (HCl)	
Gallium Arsenide (GaAs)	• Citric Acid diluted ($C_6H_8O_7$: H_2O, 1 : 1) + Hydrogen Peroxide (H_2O_2)+ Water (H_2O)	• Cl_2, CCl_4, $SiCl_4$, BCl_3, CCl_2F_2
Gold (Au)	Aqua regia, Iodine and Potassium Iodide solution	
Molybdenum (Mo)		CF_4
Organic residues and photoresist	Piranha etch: sulfuric acid (H_2SO_4) + hydrogen peroxide (H_2O_2)	O_2 (ashing)
Platinum (Pt)	Aqua regia	
Silicon (Si)	• Nitric acid (HNO_3) + hydrofluoric acid (HF) • Potassium hydroxide (KOH) • Ethylenediamine pyrocatechol (EDP) • Tetramethylammonium hydroxide (TMAH)	• CF_4, SF_6, NF_3 • Cl_2, CCl_2F_2
Silicon dioxide (SiO_2)	• Hydrofluoric acid (HF) • Buffered oxide etch [BOE]: ammonium fluoride (NH_4F) and hydrofluoric acid (HF)	CF_4, SF_6, NF_3
Silicon nitride (Si_3N_4)	• 85% Phosphoric acid (H_3PO_4) at 180 °C (Requires SiO_2 etch mask)	CF_4, SF_6, NF_3, CHF_3
Tantalum (Ta)		CF_4
Titanium (Ti)	Hydrofluoric acid (HF)	BCl_3
Titanium nitride (TiN)	• Nitric acid (HNO_3) + hydrofluoric acid (HF) • SC1 • Buffered HF (bHF)	
Tungsten (W)	• Nitric acid (HNO_3) + hydrofluoric acid (HF) • Hydrogen Peroxide (H_2O_2)	• CF_4 • SF_6

Wafer Dicing

Wafer dicing is the process by which die are separated from a wafer of semiconductor following the processing of the wafer. The dicing process can be accomplished by scribing and breaking, by mechanical sawing (normally with a machine called a *dicing saw*) or by laser cutting. All methods are typically automated to ensure precision and accuracy. Following the dicing process the individual silicon chips are encapsulated into chip carriers which are then suitable for use in building electronic devices such as computers, etc.

During dicing, wafers are typically mounted on dicing tape which has a sticky backing that holds the wafer on a thin sheet metal frame. Dicing tape has different properties depending on the dicing application. UV curable tapes are used for smaller sizes and non-UV dicing tape for larger die sizes. Once a wafer has been diced, the pieces left on the dicing tape are referred to as *die*, *dice* or *dies*. These will be packaged in a suitable package or placed directly on a printed circuit board substrate as a "bare die". The area that has been cut away are called *die streets* which are typically about 75 micrometres (0.003 inch) wide. Once a wafer has been diced, the die will stay on the dicing tape until they are extracted by die handling equipment, such as a *die bonder* or *die sorter*, further in the electronics assembly process.

The size of the die left on the tape may range from 35 mm (very large) to 0.1 mm square (very small). The die created may be any shape generated by straight lines, but they are typically rectangular or square shaped. In some cases they can be other shapes as well depending on the singulation method used. A full-cut laser dicer has the ability to cut and separate in a variety of shapes.

Semiconductor Device Fabrication

Integrated circuits are crucial in electronic and electrical devices. The creation of integrated circuits is enabled by semiconductor device fabrication. It involves a sequence of chemical and photolithographic processing for the gradual creation of electronic circuits on a pure semiconducting material wafer. Silicon is an important material for the process, but different compound semiconductors can also be used for specialized applications. The whole manufacturing process, from start to packaging and shipment, is a long one that is performed in specialized facilities.

Wafers

A typical wafer is made out of extremely pure silicon that is grown into mono-crystalline cylindrical ingots (boules) up to 300 mm (slightly less than 12 inches) in diameter using the Czochralski process. These ingots are then sliced into wafers about 0.75 mm thick and polished to obtain a very regular and flat surface.

Processing

In semiconductor device fabrication, the various processing steps fall into four general categories: deposition, removal, patterning, and modification of electrical properties.

- *Deposition* is any process that grows, coats, or otherwise transfers a material onto the wafer. Available technologies include physical vapor deposition (PVD), chemical vapor deposition (CVD), electrochemical deposition (ECD), molecular beam epitaxy (MBE) and more recently, atomic layer deposition (ALD) among others.
- *Removal* is any process that removes material from the wafer; examples include etch processes (either wet or dry) and chemical-mechanical planarization (CMP).
- *Patterning* is the shaping or altering of deposited materials, and is generally referred to as lithography. For example, in conventional lithography, the wafer is coated with a chemical called a *photoresist*; then, a machine called a *stepper* focuses, aligns, and moves a mask, exposing select portions of the wafer below to short wavelength light; the exposed regions are washed away by a developer solution. After etching or other processing, the remaining photoresist is removed by plasma ashing.
- *Modification of electrical properties* has historically entailed doping transistor *sources* and *drains* (originally by diffusion furnaces and later by ion implantation). These doping processes are followed by furnace annealing or, in advanced devices, by rapid thermal annealing (RTA); annealing serves to activate the implanted dopants. Modification of electrical properties now also extends to the reduction of a material's dielectric constant in low-k insulators via exposure to ultraviolet light in UV processing (UVP). Modification is frequently achieved by oxidation, which can be carried out to create semiconductor-insulator junctions, such as in the local oxidation of silicon (LOCOS) to fabricate field effect transistors.

Modern chips have up to eleven metal levels produced in over 300 sequenced processing steps.

Front-end-of-line (FEOL) Processing

FEOL processing refers to the formation of the transistors directly in the silicon. The raw wafer is engineered by the growth of an ultrapure, virtually defect-free silicon layer through epitaxy. In the most advanced logic devices, *prior* to the silicon epitaxy step, tricks are performed to improve the performance of the transistors to be built. One method involves introducing a *straining step* wherein a silicon variant such as silicon-germanium (SiGe) is deposited. Once the epitaxial silicon is deposited, the crystal lattice becomes stretched somewhat, resulting in improved electronic mobility. Another method, called *silicon on insulator* technology involves the insertion of an insulating layer between the raw silicon wafer and the thin layer of subsequent silicon epitaxy. This method results in the creation of transistors with reduced parasitic effects.

Gate Oxide and Implants

Front-end surface engineering is followed by growth of the gate dielectric (traditionally silicon dioxide), patterning of the gate, patterning of the source and drain regions, and subsequent implantation or diffusion of dopants to obtain the desired complementary electrical properties. In dynamic random-access memory (DRAM) devices, storage capacitors are also fabricated at this time, typically stacked above the access transistor (the now defunct DRAM manufacturer Qimonda implemented these capacitors with trenches etched deep into the silicon surface).

Back-end-of-line (BEOL) Processing

Metal Layers

Once the various semiconductor devices have been created, they must be interconnected to form the desired electrical circuits. This occurs in a series of wafer processing steps collectively referred to as BEOL. BEOL processing involves creating metal interconnecting wires that are isolated by dielectric layers. The insulating material has traditionally been a form of SiO_2 or a silicate glass, but recently new low dielectric constant materials are being used (such as silicon oxycarbide), typically providing dielectric constants around 2.7 (compared to 3.82 for SiO_2), although materials with constants as low as 2.2 are being offered to chipmakers.

Interconnect

Synthetic detail of a standard cell through four layers of planarized copper interconnect, down to the polysilicon (pink), wells (greyish) and substrate (green).

Historically, the metal wires have been composed of aluminum. In this approach to wiring (often called *subtractive aluminum*), blanket films of aluminum are deposited first, patterned, and then etched, leaving isolated wires. Dielectric material is then deposited over the exposed wires. The various metal layers are interconnected by etching holes (called *"vias")* in the insulating material and then depositing tungsten in them with a CVD technique; this approach is still used in the fabrication of many memory chips such as dynamic random-access memory (DRAM), because the number of interconnect levels is small (currently no more than four).

More recently, as the number of interconnect levels for logic has substantially increased due to the large number of transistors that are now interconnected in a modern microprocessor, the timing delay in the wiring has become so significant as to prompt a change in wiring material (from aluminum to copper interconnect layer) and a change in dielectric material (from silicon dioxides to newer low-K insulators). This performance enhancement also comes at a reduced cost via damascene processing, which eliminates processing steps. As the number of interconnect levels increases, planarization of the previous layers is required to ensure a flat surface prior to subsequent lithography. Without it, the levels would become increasingly crooked, extending outside the depth of focus of available lithography, and thus interfering with the ability to pattern. CMP (chemical-mechanical planarization) is the primary processing method to achieve such planarization, although dry *etch back* is still sometimes employed when the number of interconnect levels is no more than three.

Wafer Test

The highly serialized nature of wafer processing has increased the demand for metrology in between the various processing steps. For example, thin film metrology based on ellipsometry or reflectometry is used to tightly control the thickness of gate oxide, as well as the thickness, refractive index and extinction coefficient of photoresist and other coatings. Wafer test metrology equipment is used to verify that the wafers haven't been damaged by previous processing steps up until testing; if too many dies on one wafer have failed, the entire wafer is scrapped to avoid the costs of further processing. Virtual metrology has been used to predict wafer properties based on statistical methods without performing the physical measurement itself.

Device Test

Once the front-end process has been completed, the semiconductor devices are subjected to a variety of electrical tests to determine if they function properly. The proportion of devices on the wafer found to perform properly is referred to as the yield. Manufacturers are typically secretive about their yields, but it can be as low as 30%. Process variation is one among many reasons for low yield.

The fab tests the chips on the wafer with an electronic tester that presses tiny probes against the chip. The machine marks each bad chip with a drop of dye. Currently, electronic dye marking is possible if wafer test data is logged into a central computer database and chips are "binned" (i.e. sorted into virtual bins) according to the predetermined test limits. The resulting binning data can be graphed, or logged, on a wafer map to trace manufacturing defects and mark bad chips. This map can also be used during wafer assembly and packaging.

Chips are also tested again after packaging, as the bond wires may be missing, or analog performance may be altered by the package. This is referred to as the "final test".

Usually, the fab charges for testing time, with prices in the order of cents per second. Testing times vary from a few milliseconds to a couple of seconds, and the test software is optimized for reduced testing time. Multiple chip (multi-site) testing is also possible, because many testers have the resources to perform most or all of the tests in parallel.

Chips are often designed with "testability features" such as scan chains or a "built-in self-test" to speed testing, and reduce testing costs. In certain designs that use specialized analog fab processes, wafers are also laser-trimmed during the testing, in order to achieve tightly-distributed resistance values as specified by the design.

Good designs try to test and statistically manage *corners* (extremes of silicon behavior caused by a high operating temperature combined with the extremes of fab processing steps). Most designs cope with at least 64 corners.

Die Preparation

Once tested, a wafer is typically reduced in thickness before the wafer is scored and then broken into individual dice, a process known as wafer dicing. Only the good, unmarked chips are packaged.

Packaging

Plastic or ceramic packaging involves mounting the die, connecting the die pads to the pins on the package, and sealing the die. Tiny wires are used to connect the pads to the pins. In the old days, wires were attached by hand, but now specialized machines perform the task. Traditionally, these wires have been composed of gold, leading to a lead frame of solder-plated copper; lead is poisonous, so lead-free "lead frames" are now mandated by RoHS.

Chip scale package (CSP) is another packaging technology. A plastic dual in-line package, like most packages, is many times larger than the actual die hidden inside, whereas CSP chips are nearly the size of the die; a CSP can be constructed for each die *before* the wafer is diced.

The packaged chips are retested to ensure that they were not damaged during packaging and that the die-to-pin interconnect operation was performed correctly. A laser then etches the chip's name and numbers on the package.

List of Steps

This is a list of processing techniques that are employed numerous times throughout the construction of a modern electronic device; this list does not necessarily imply a specific order.

- Wafer processing
  - Wet cleans
    - Cleaning by solvents such as acetone, trichloroethylene
    - Piranha solution
    - RCA clean
  - Photolithography
  - Ion implantation (in which dopants are embedded in the wafer creating regions of increased (or decreased) conductivity)
  - Dry etching
  - Wet etching
  - Plasma ashing
  - Thermal treatments
    - Rapid thermal anneal
    - Furnace anneals
    - Thermal oxidation
  - Chemical vapor deposition (CVD)
  - Physical vapor deposition (PVD)
  - Molecular beam epitaxy (MBE)
  - Electrochemical deposition (ECD).
  - Chemical-mechanical planarization (CMP)
  - Wafer testing (where the electrical performance is verified)
  - Wafer backgrinding (to reduce the thickness of the wafer so the resulting chip can be put into a thin device like a smartcard or PCMCIA card.)
- Die preparation
  - Wafer mounting
  - Die cutting

- IC packaging
    - Die attachment
    - IC bonding
        - Wire bonding
        - Thermosonic bonding
        - Flip chip
        - Wafer bonding
        - Tape Automated Bonding (TAB)
    - IC encapsulation
        - Baking
        - Plating
        - Lasermarking
        - Trim and form
- IC testing

Hazardous Materials

Many toxic materials are used in the fabrication process. These include:

- poisonous elemental dopants, such as arsenic, antimony, and phosphorus.
- poisonous compounds, such as arsine, phosphine, and silane.
- highly reactive liquids, such as hydrogen peroxide, fuming nitric acid, sulfuric acid, and hydrofluoric acid.

It is vital that workers should not be directly exposed to these dangerous substances. The high degree of automation common in the IC fabrication industry helps to reduce the risks of exposure. Most fabrication facilities employ exhaust management systems, such as wet scrubbers, combustors, heated absorber cartridges, etc., to control the risk to workers and to the environment.

Physical Vapor Deposition

Physical vapour deposition (PVD) is fundamentally a vaporisation coating technique, involving transfer of material on an atomic level. It is an alternative process to electroplating.

The process is similar to chemical vapour deposition (CVD) except that the raw materials/precursors, i.e. the material that is going to be deposited starts out in solid form, whereas in CVD, the precursors are introduced to the reaction chamber in the gaseous state.

It incorporates processes such as sputter coating and pulsed laser deposition (PLD).

Steps of Physical Vapour Deposition

PVD processes are carried out under vacuum conditions. The process involved four steps:

- Evaporation
- Transportation
- Reaction
- Deposition.

Evaporation

During this stage, a target, consisting of the material to be deposited is bombarded by a high ebergy source suchg as a beam of electrons or ions. This dislodges atoms from the surface of the target, 'vaporising' them.

Transport

This process simply consists of the movement of 'vaporised' atoms from the target to the substrate to be coated and will generally be a straight line affair.

Reaction

In some cases, coatings will consist of metal oxides, nitrides, carbides and other such materials. In these cases, the target will consist of the metal. The atoms of metal will then react with the appropriate gas during the transport stage. For the above examples, the reactive gases may be oxygen, nitrogen and methane.

In instances where the coating consists of the target material alone, this step would not be part of the process.

Deposition

This is the process of coating build up on the substrate surface.

Depending on the actual process, some reactions between target materials and the reactive gases may also take place at the substrate surface simultaneously with the deposition process.

Uses of PVD Coatings

PVD coatings are deposited for numerous reasons. Some of the main ones are:

- Improved hardness and wear resistance
- Reduced friction
- Improved oxidation resistance.

The use of such coatings is aimed at improving efficiency through improved performance and longer component life. They may also allow coated components to operate in environments that the uncoated component would not otherwise have been able to perform.

Advantages of the Physical Vapour Deposition Process

- Materials can be deposited with improved properties compared to the substrate material
- Almost any type of inorganic material can be used as well as some kinds of organic materials
- The process is more environmentally friendly than processes such as electroplating.

Disadvantages of the Physical Vapour Deposition Process

- It is a line of sight technique meaning that it is extremely difficult to coat undercuts and similar surface features
- High capital cost
- Some processes operate at high vacuums and temperatures requiring skilled operators
- Processes requiring large amounts of heat require appropriate cooling systems
- The rate of coating deposition is usually quite slow.

Applications

PVD coatings are generally used to improve hardness, wear resistance and oxidation resistance. Thus, such coatings use in a wide range of applications such as:

- Aerospace
- Automotive
- Surgical/Medical

- Dies and moulds for all manner of material processing
- Cutting tools
- Fire arms.

Chemical Vapor Deposition

Chemical vapor deposition (CVD) is one of the most common processes used to coat almost any metallic or ceramic compound, including elements, metals and their alloys and intermetallic compounds. The CVD process involves depositing a solid material from a gaseous phase; this is achieved by means of a chemical reaction between volatile precursors and the surface of the materials to be coated. As the precursor gases pass over the surface of the heated substrate, the resulting chemical reaction forms a solid phase which is deposited onto the substrate. The substrate temperature is critical and can influence the occurrence of different reactions.

Chemical vapor deposition (CVD) is a technology used for forming solid powders in the gas phase or depositing films on substrates from precursors in the gas phase. Many precursors are liquids or solids under ordinary conditions, and they are first vaporized for use in CVD. Based on the process mainly being physical or chemical, vapor deposition has been divided into two kinds, namely physical vapor deposition (PVD) and chemical vapor deposition. Vacuum vaporization, spattering, ion plating, etc. usually belong to PVD; and vapor depositions based on chemical reactions or plasma-enhanced chemical reactions are called CVD or plasma-enhanced chemical vapor deposition (PECVD or PCVD).

Rapid Thermal Processing

Rapid thermal processing (or RTP) refers to a semiconductor manufacturing process which heats silicon wafers to high temperatures (up to 1200°C or greater) on a timescale of several seconds or less. Such rapid heating rates are attained by high intensity lamps or laser process. These processes are used for a wide variety of applications in semiconductor manufacturing including dopant activation, thermal oxidation, metal reflow and chemical vapor deposition.

Temperature Control

One of the key challenges in rapid thermal processing is accurate measurement and control of the wafer temperature. Monitoring the ambient with a thermocouple is not feasible, in that the high temperature ramp rates prevent the wafer from coming to thermal equilibrium with the process chamber. One temperature control strategy involves in situ pyrometry to effect real time control.

Bulk Micromachining

Bulk micromachining (BMM) is a set of processes that enable the 3D sculpting of various materials (mainly silicon) to make small structures that serve as components for MEMS devices. Bulk micromachining builds mechanical elements by starting with a bulk material, and then etching away unwanted parts, and being left with useful mechanical devices. Typically, the substrate is photo-patterned, leaving a protective layer on the parts of the substrate that are to be retained. The substrate is then etched using a wet chemical etch process, dry chemical etch process, reactive plasma etch process, or photoetch process. The etch process "eats away" any exposed substrate material on the substrate. Bulk micromachining is a relatively simple and inexpensive fabrication technology to fabricate a micromachined device.

References

- Dixit, U.S.; Das, R. (2012). "Chapter 15: Microextrusion". In Jain, V.K. Micromanufacturing Processes. CRC Press. pp. 263–282. ISBN 9781439852903.
- 81-what-is-thin-film-deposition: semicore.com, Retrieved 14 July 2018
- Razali, A.R.; Qin, Y. (2013). "A review on micro-manufacturing, micro-forming and their key issues". Procedia Engineering. 53 (2013): 665–672. doi:10.1016/j.proeng.2013.02.086.
- Tell-me-what-is-wafer-dicing: escomponents.com, Retrieved 29 March 2018
- Advanced Manufacturing Processes Laboratory (2015). "Process Analysis and Variation Control in Micro-stamping". Northwestern University. Retrieved 18 March 2016.
- Chemical-vapor-deposition, materials-science: sciencedirect.com, Retrieved 16 April 2018
- Fu, M.W.; Chan, W.L. (2013). "A review on the state-of-the-art microforming technologies". International Journal of Advanced Manufacturing Technology. 67 (9): 2411–2437. doi:10.1007/s00170-012-4661-7
- Rapid-thermal-processing-equipment: agsemiconductor.com, Retrieved 27 May 2018
- Lécuyer, Christophe (2010). Makers of the Microchip: A Documentary History of Fairchild Semiconductor. The MIT Press. ISBN 978-0262014243.

3

Important Electrical and Mechanical Concepts

Some of the fundamental concepts central to the understanding of microfabrication and MEMS technology are semiconductor conductivity, resistivity, beam theory and beam deflection, etc. The aim of this chapter is to provide a comprehensive understanding of MEMS and includes significant electrical and mechanical concepts pertaining to electrical resistance, conductance, stress and strain, etc.

Semiconductor Conductivity

The conductivity of a semiconductor is given by:

$$\sigma = q(\mu_n n + \mu_p p)$$

where μ_n and μ_p refer to the mobilities of the electrons and holes, and n and p refer to the density of electrons and holes, respectively. A doped semiconductor, majority carriers greatly outnumber minority carriers, so that above equation can be reduced to a single term involving the majority carrier.

Effects of Temperature and Doping on Mobility of a Semiconductor

Conductivity of a material is determined by two factors: the concentration of free carriers available to conduct current and their mobility. In a semiconductor, both mobility and carrier concentration are temperature dependent. Thus, it is important to view the conductivity as a function of temperature which is expressed by:

$$\sigma = q\left[\mu_n(T)n(T) + \mu_p(T)p(T)\right]$$

There are two basic types of scattering mechanisms that influence the mobility of electrons and holes: lattice scattering and impurity scattering. Lattice vibrations cause the mobility to decrease with increasing temperature.

However, the mobility of the carriers in a semiconductor is also influenced by the presence of charged impurities. Impurity scattering is caused by crystal defects such as ionized impurities. At lower temperatures, carriers move more slowly, so there is more time for them to interact with charged impurities. As a result, as the temperature decreases, impurity scattering increases and the mobility decreases. This is just the opposite of the effect of lattice scattering.

The total mobility then is the sum of the lattice-scattering mobility and the impurity-scattering mobility. Figure shows how the total mobility has a temperature at which it is a maximum. The approximate temperature dependence of mobility due to lattice scattering is $T^{+3/2}$, while the temperature dependence of mobility due to impurity scattering is $T^{+3/2}$. In practice, impurity scattering is typically only seen at very low temperatures. In the temperature range we will measure, only the influence of lattice scattering will be expected.

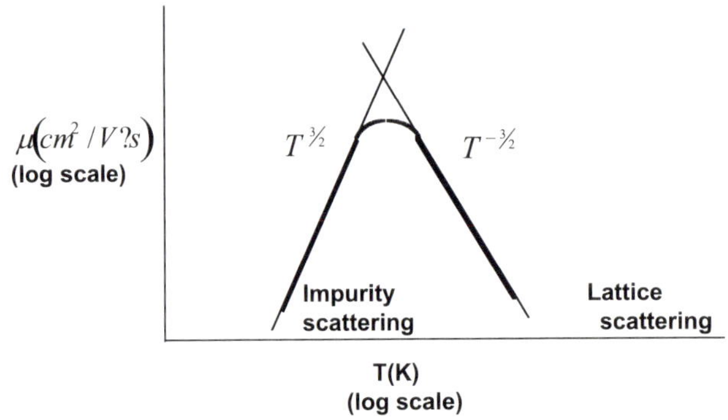

Figure: Approximate temperature dependence of mobility with both lattice and impurity scattering

Temperature Dependence of Carrier Concentration

The carrier concentration in a semiconductor is also affected by temperature. The intrinsic carrier concentration is governed by:

$$n_i(T) = 2\left[\frac{2\pi kT}{h^2}\right]^{\frac{3}{2}}\left(m_n^* m_p^*\right)^{\frac{3}{4}} \exp\left[\frac{E_g}{2kT}\right]$$

where the exponential temperature dependence dominates $n_i(T)$. To determine the total carrier concentration, we must also consider space-charge neutrality:

$$n(T) = N_D^+(T) - N_A^- + \frac{n_i^2(T)}{n(T)} \quad \text{and:} \quad p(T) = N_A^-(T) - N_D^+(T) + \frac{n_i^2(T)}{p(T)}$$

For a doped semiconductor, the temperature dependence of electron concentration can be seen in figure. At very low temperatures, negligible intrinsic electron-hole-pairs exist (n_i is very small), and the donor electrons are bound to the donor atoms. This is known as the ionization (or freeze-out) region. As the temperature is raised, increased ionization occurs and at about 100K all of the donor atoms are ionized, at which point the carrier concentration is determined by doping. The region where every available dopant has been ionized is called the extrinsic region. In this region, an increase in temperature produces no increase in carrier concentration. Referring to equation above, this is the region where $N_D^+(T) = N_D$, $N_A^-(T) = N_A$, and $n_i(T) \ll |N_D - N_A|$. At high temperatures, the thermally generated intrinsic carriers outnumber the dopants $(n_i > |N_D - N_A|)$. In this intrinsic region, carrier concentration increases with temperature as shown in equation $n_i(T) = 2\left[\frac{2\pi kT}{h^2}\right]^{\frac{3}{2}} (m_n^* m_p^*)^{\frac{3}{4}} \exp\left[\frac{E_g}{2kT}\right]$ because n_i has become the dominant term of Equation $\left(n(T) = N_D^+(T) - N_A^- + \frac{n_i^2(T)}{n(T)}\right.$ and: $\left. p(T) = N_A^-(T) - N_D^+(T) + \frac{n_i^2(T)}{p(T)}\right)$.

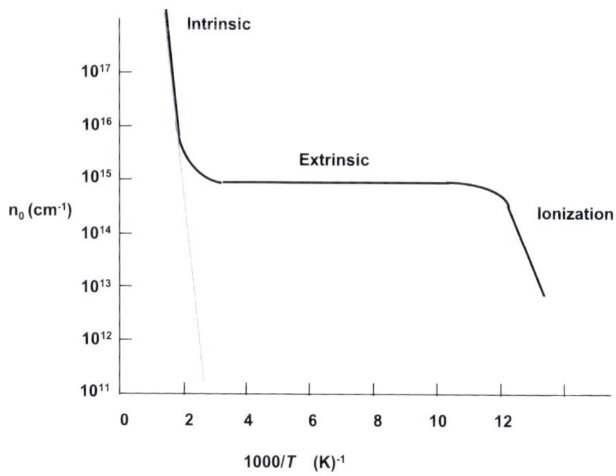

Figure: Carrier concentration vs. reciprocal temperature for silicon doped with 10^{15} donors/cm^3

Temperature Dependence of Conductivity for a Semiconductor

Remember that Equation ($\sigma = q\left[\mu_n(T)n(T) + \mu_p(T)p(T)\right]$) showed that conductivity depends on both carrier concentration and mobility, so there are a variety of possible temperature dependencies for conductivity. For instance, at fairly low temperatures (less than 200K), the dominant scattering mechanism might be impurity scattering $\left(\mu \, \alpha \, T^{3/2}\right)$ while the carrier concentration is determined by extrinsic doping $\left(n = N_{D^+}\right)$ therefore, conductivity would be seen to increase with temperature $\left(\sigma \, \alpha \, T^{3/2}\right)$. Other possibilities, depending on the material, doping, and temperature will show different temperature dependence of conductivity. One particularly interesting case occurs at high temperatures (above 400K or higher) when carrier concentration is intrinsic (Equation $\left(n(T) = N_D^+(T) - N_A^- + \frac{n_i^2(T)}{n(T)}\right.$ and: $\left. p(T) = N_A^-(T) - N_D^+(T) + \frac{n_i^2(T)}{p(T)}\right)$) and mobility is dominated by lattice scattering $\left(\mu \, \alpha \, T^{-3/2}\right)$. In such cases, the conductivity can easily be shown to vary with temperature as: $\sigma \propto \exp\left(\frac{-E_g}{2kT}\right)$

In this case, conductivity depends only on the semiconductor bandgap and the temperature. In this temperature range, measured conductivity data can be used to determine the semiconductor bandgap energy, E_g.

Electrical Resistance and Conductance

Resistance

As its name implies, a resistor is an electronic component that resists the flow of electric current in a circuit. Electrical resistance is analogous to friction in a mechanical system. They both convert energy to heat and dissipate it to the surrounding environment, so electrical resistance can sometimes be thought of as a braking or damping mechanism in a circuit.

In metals such as silver or copper, which have high electrical conductivity and therefore low resistivity, electrons are able to skip freely from the conduction band of one atom to the next, encountering little resistance. However, in a material such as carbon, electrons encounter numerous collisions that make it more difficult for them to move through the material, according to Serif Uran, a professor of physics at Pittsburg State University. Insulators such as glass have extremely high resistivity, with virtually no spaces in their conduction bands that would allow electrons to move through them.

The electrical resistance of a circuit component is defined as the ratio of the applied voltage to the electric current that flows through it, according to HyperPhysics. The standard unit for resistance is the ohm, which is named after German physicist Georg Simon Ohm. It is defined as the resistance in a circuit with a current of one ampere at one volt. Resistance can be calculated using Ohm's Law, which states that resistance equals voltage divided by current, or $R = V/I$, where R is resistance, V is voltage and I is current. Ohm's Law is more commonly written as the equivalent expression $V = IR$. One way to understand Ohm's Law is to hold one of these variables constant, change the value of another variable, and watch what happens to the third variable. For instance, if we keep voltage constant and increase the resistance, the current must decrease. If we keep the resistance constant and increase the voltage, the current must increase.

Resistors are generally classified as either fixed or variable. Fixed-value resistors are simple passive components that always have the same resistance within their prescribed current and voltage limits. They are available in a wide range of resistance values from less than 1 ohm to several million ohms with tolerances ranging from plus or minus 0.1 percent to plus or minus 10 percent. Resistors are also classified by the maximum voltage they can tolerate as well as the maximum amount of power they can dissipate.

The resistance of a simple resistor can be calculated as $R = \rho L/A$, where R is resistance, L is its length, A is its cross-sectional area and ρ is resistivity, which is an inherent property of the material. Resistivity is the reciprocal of conductivity σ, *i.e.*,$\rho = 1/\sigma$. All other things being equal, a resistor that is twice as long will have twice the resistance, and one with twice the cross-sectional areal will have half the resistance. Also, material with higher resistivity will result in proportionally greater resistance.

Variable resistors are simple electro-mechanical devices, such as volume controls and dimmer switches, which increase the effective length of a resistor by turning a knob or moving a slide control. Strain gauges are resistors in which resistance changes with strain. Strain occurs when an object is stretched or compressed. A thermistor is a temperature sensor. It changes resistance when an increase in temperature excites electrons, making them available to conduct current, thus reducing the resistivity of the material. A piezoresistor changes its resistivity in response to a change in strain, which causes more or fewer electrons to be available to carry charge.

Conductance

Conductance is the measure of how easily electricity flows along a certain path through an electrical element, and since electricity is so often explained in terms of opposites, conductance is considered the opposite of resistance. In terms of resistance and conductance, the reciprocal relationship between the two can be expressed through the following equation: $R = 1/G$, $G=1/R$; where R equals resistance and G equals conduction.

Another way to represent this is: $W=1/S$, $S=1/W$, where W (the Greek letter omega) represents resistance and S represents Siemens, ergo the measure of conductance. In addition, Siemens can be measured by comparing them to their equivalent of one ampere (A) per volt (V).

In other words, when a current of one ampere (1A) passes through a component across which a voltage of one volt (1V) exists, then the conductance of that component is one Siemens (1S). This can be expressed through the equation: $G = I/E$, where G represents conductance and E is the voltage across the component (expressed in volts).

The temperature of the material is definitely a factor, but assuming a constant temperature, the conductance of a material can be calculated.

The SI (International System) derived unit of conductance is known as the Siemens, named after the German inventor and industrialist Ernst Werner von Siemens. Since conductance is the opposite of resistance, it is usually expressed as the reciprocal of one ohm – a unit of electrical resistance named after George Simon Ohm – or one mho (ohm spelt backwards).

Recently, this term was re-designated to Siemens, expressed by the notational symbol S. The factors that affect the magnitude of resistance are exactly the same for conductance, but they affect conductance in the opposite manner. Therefore, conductance is directly proportional to area, and inversely proportional to the length of the material.

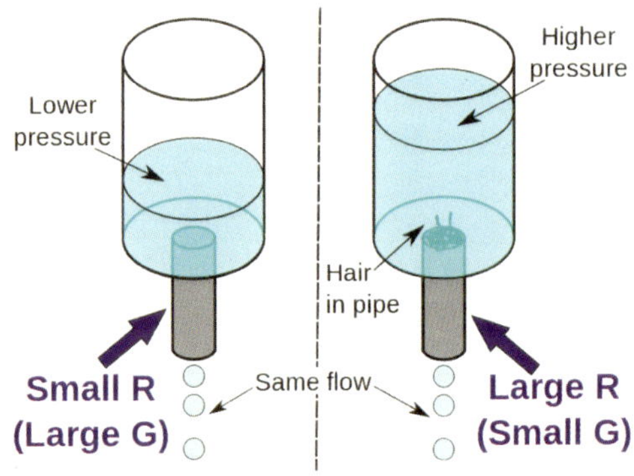

The hydraulic analogy compares electric current flowing through circuits to water flowing through pipes. When a pipe (left) is filled with hair (right), it takes a larger pressure to achieve the same flow of water. Pushing electric current through a large resistance is like pushing water through a pipe clogged with hair: It requires a larger push (electromotive force) to drive the same flow (electric current).

In the hydraulic analogy, current flowing through a wire (or resistor) is like water flowing through a pipe, and the voltage drop across the wire is like the pressure drop that pushes water through the pipe. Conductance is proportional to how much flow occurs for a given pressure, and resistance is proportional to how much pressure is required to achieve a given flow. (Conductance and resistance are reciprocals.)

The voltage *drop* (i.e., difference between voltages on one side of the resistor and the other), not the voltage itself, provides the driving force pushing current through a resistor. In hydraulics, it is similar: The pressure *difference* between two sides of a pipe, not the pressure itself, determines the flow through it. For example, there may be a large water pressure above the pipe, which tries to push water down through the pipe. But there may be an equally large water pressure below the pipe, which tries to push water back up through the pipe. If these pressures are equal, no water flows. (In the image at right, the water pressure below the pipe is zero.)

The resistance and conductance of a wire, resistor, or other element is mostly determined by two properties:

- Geometry (shape), and
- Material

Geometry is important because it is more difficult to push water through a long, narrow pipe than a wide, short pipe. In the same way, a long, thin copper wire has higher resistance (lower conductance) than a short, thick copper wire.

Materials are important as well. A pipe filled with hair restricts the flow of water more than a clean pipe of the same shape and size. Similarly, electrons can flow freely and easily through a copper wire, but cannot flow as easily through a steel wire of the same shape and size, and they essentially cannot flow at all through an insulator like rubber, regardless of its shape. The difference between copper, steel, and rubber is related to their microscopic structure and electron configuration, and is quantified by a property called resistivity.

In addition to geometry and material, there are various other factors that influence resistance and conductance, such as temperature.

Conductors and Resistors

A 6.5 MΩ resistor, as identified by its electronic color code (blue–green–black-yellow-red). An ohmmeter could be used to verify this value.

Substances in which electricity can flow are called conductors. A piece of conducting material of a particular resistance meant for use in a circuit is called a resistor. Conductors are made of high-conductivity materials such as metals, in particular copper and aluminium. Resistors, on the other hand, are made of a wide variety of materials depending on factors such as the desired resistance, amount of energy that it needs to dissipate, precision, and costs.

Ohm's Law

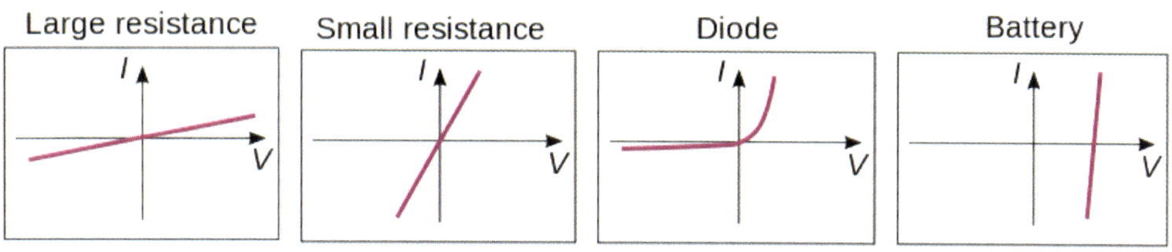

The current-voltage characteristics of four devices: Two resistors, a diode, and a battery. The horizontal axis is voltage drop, the vertical axis is current. Ohm's law is satisfied when the graph is a straight line through the origin. Therefore, the two resistors are *ohmic*, but the diode and battery are not.

For many materials, the current I through the material is proportional to the voltage V applied across it:

$$I \propto V$$

Therefore, the resistance and conductance of objects or electronic components made of these materials is constant. This relationship is called Ohm's law, and materials which obey it are called *ohmic* materials. Examples of ohmic components are wires and resistors. The current-voltage (IV) graph of an ohmic device consists of a straight line through the origin with positive slope.

Other components and materials used in electronics do not obey Ohm's law; the current is not proportional to the voltage, so the resistance varies with the voltage and current through them. These are called *nonlinear* or *nonohmic*. Examples include diodes and fluorescent lamps. The IV curve of a nonohmic device is a curved line.

Relation to Resistivity and Conductivity

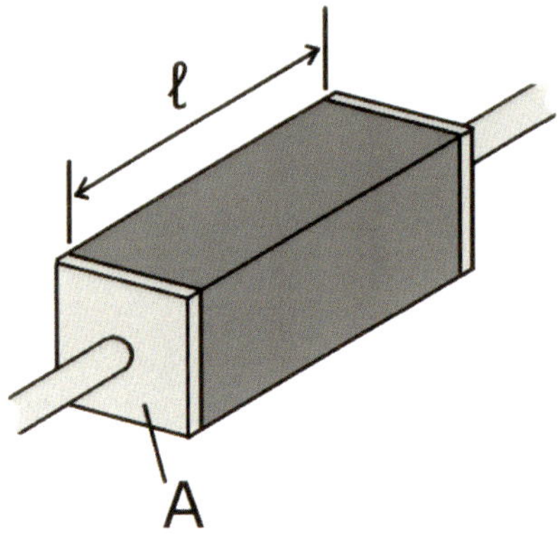

A piece of resistive material with electrical contacts on both ends.

The resistance of a given object depends primarily on two factors: What material it is made of, and its shape. For a given material, the resistance is inversely proportional to the cross-sectional area; for example, a thick copper wire has lower resistance than an otherwise-identical thin copper wire. Also, for a given material, the resistance is proportional to the length; for example, a long copper wire has higher resistance than an otherwise-identical short copper wire. The resistance R and conductance G of a conductor of uniform cross section, therefore, can be computed as:

$$R = \rho\frac{\ell}{A},$$

$$G = \sigma\frac{A}{\ell}.$$

where ℓ is the length of the conductor, measured in metres [m], A is the cross-sectional area of the conductor measured in square metres [m^2], σ (sigma) is the electrical conductivity measured in siemens per meter ($S{\cdot}m^{-1}$), and ρ (rho) is the electrical resistivity (also called *specific electrical resistance*) of the material, measured in ohm-metres $(\Omega{\cdot}m)$. The resistivity and conductivity are proportionality constants, and therefore depend only on the material the wire is made of, not the geometry of the wire. Resistivity and conductivity are reciprocals: $\rho = 1/\sigma$. Resistivity is a measure of the material's ability to oppose electric current.

This formula is not exact, as it assumes the current density is totally uniform in the conductor, which is not always true in practical situations. However, this formula still provides a good approximation for long thin conductors such as wires.

Another situation for which this formula is not exact is with alternating current (AC), because the skin effect inhibits current flow near the center of the conductor. For this reason, the *geometrical* cross-section is different from the *effective* cross-section in which current actually flows, so resistance is higher than expected. Similarly, if two conductors near each other carry AC current, their resistances increase due to the proximity effect. At commercial power frequency, these effects are significant for large conductors carrying large currents, such as busbars in an electrical substation, or large power cables carrying more than a few hundred amperes.

Determination of Resistivity

The resistivity of different materials varies by an enormous amount. For example, the conductivity of teflon is about 10^{30} times lower than the conductivity of copper. Why is there such a difference? Loosely speaking, a metal has large numbers of "delocalized" electrons that are not stuck in any one place, but free to move across large distances, whereas in an insulator (like teflon), each electron is tightly bound to a single molecule, and a great force is required to pull it away. Semiconductors lie between these two extremes.

Resistivity varies with temperature. In semiconductors, resistivity also changes when exposed to light.

Measuring Resistance

An instrument for measuring resistance is called an ohmmeter. Simple ohmmeters cannot measure low resistances accurately because the resistance of their measuring leads causes a voltage drop that interferes with the measurement, so more accurate devices use four-terminal sensing.

Typical Resistances

Component	Resistance (Ω)
1 meter of copper wire with 1 mm diameter	0.02
1 km overhead power line (typical)	0.03
AA battery (typical internal resistance)	0.1
Incandescent light bulb filament (typical)	200–1000
Human body	1000 to 100,000

Static and Differential Resistance

Many electrical elements, such as diodes and batteries do *not* satisfy Ohm's law. These are called *non-ohmic* or *non-linear*, and their I–V curves are *not* straight lines through the origin.

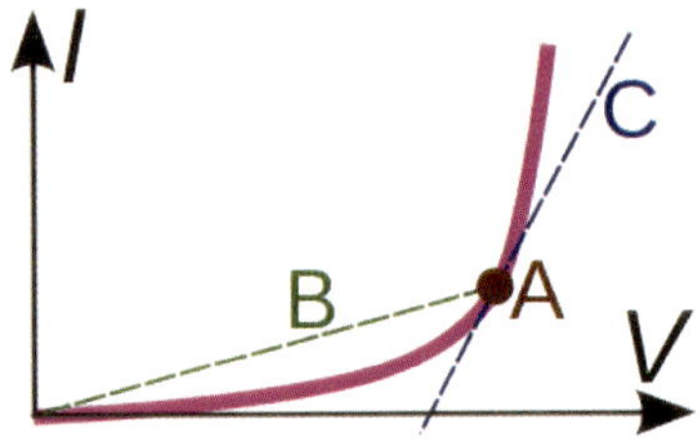

The IV curve of a non-ohmic device (purple). The static resistance at point *A* is the inverse slope of line *B*through the origin. The differential resistance at *A* is the inverse slope of tangent line *C*.

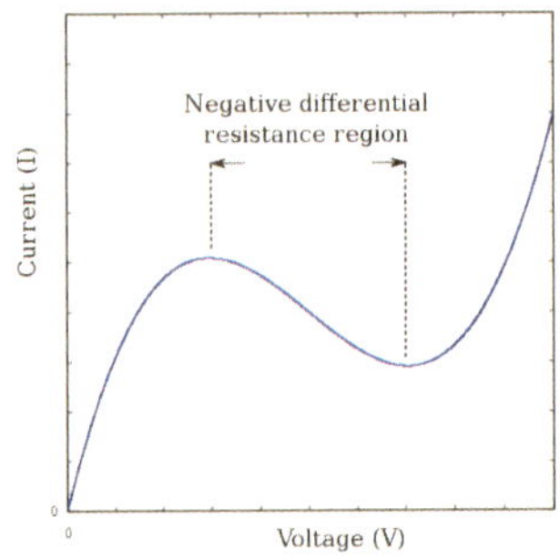

The IV curve of a component with negative differential resistance, an unusual phenomenon where the IV curve is non-monotonic.

Resistance and conductance can still be defined for non-ohmic elements. However, unlike ohmic resistance, non-linear resistance is not constant but varies with the voltage or current through the device; i.e., its operating point. There are two types of resistance:

- Static resistance (also called *chordal* or *DC resistance*) – This corresponds to the usual definition of resistance; the voltage divided by the current,

$$R_{static} = \frac{V}{T}.$$

It is the slope of the line (chord) from the origin through the point on the curve. Static resistance determines the power dissipation in an electrical component. Points on the *IV* curve located in the 2nd or 4th quadrants, for which the slope of the chordal line is negative, have *negative static resistance*. Passive devices, which have no source of energy, cannot have negative static resistance. However active devices such as transistors or op-amps can synthesize negative static resistance with feedback, and it is used in some circuits such as gyrators.

- Differential resistance (also called *dynamic, incremental* or *small signal resistance*) – Differential resistance is the derivative of the voltage with respect to the current; the slope of the *IV* curve at a point,

$$R_{\text{diff}} = \frac{dV}{dT} R_{static}.$$

If the *IV* curve is nonmonotonic (with peaks and troughs), the curve has a negative slope in some regions—so in these regions the device has *negative differential resistance*. Devices with negative differential resistance can amplify a signal applied to them, and are used to make amplifiers and oscillators. These include tunnel diodes, Gunn diodes, IMPATT diodes, magnetron tubes, and unijunction transistors.

AC Circuits

Impedance and Admittance

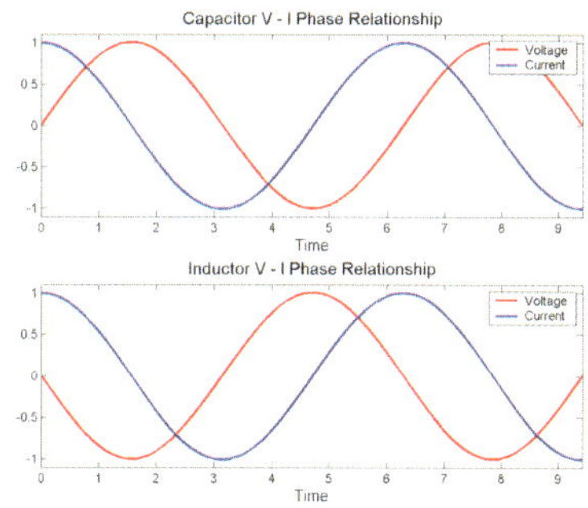

The voltage (red) and current (blue) versus time (horizontal axis) for a capacitor (top) and inductor(bottom). Since the amplitude of the current and voltage sinusoids are the same, the absolute value of impedance is 1 for both the capacitor and the inductor (in whatever units the graph is using). On the other hand, the phase difference between current and voltage is −90° for the capacitor; therefore, the complex phase of the impedance of the capacitor is −90°. Similarly, the phase difference between current and voltage is +90° for the inductor; therefore, the complex phase of the impedance of the inductor is +90°.

When an alternating current flows through a circuit, the relation between current and voltage across a circuit element is characterized not only by the ratio of their magnitudes, but also the difference in their phases. For example, in an ideal resistor, the moment when the voltage reaches its maximum, the current also reaches its maximum (current and voltage are oscillating in phase). But for a capacitor or inductor, the maximum current flow occurs as the voltage passes through zero and vice versa (current and voltage are oscillating 90° out of phase). Complex numbers are used to keep track of both the phase and magnitude of current and voltage:

$$V(t) = \mathrm{Re}(V_0 e^{j\omega t}), \quad I(t) = \mathrm{Re}(I_0 e^{j\omega t}), \quad Z = \frac{V_0}{I_0}, \quad Y = \frac{I_0}{V_0}$$

where:

- t is time,
- $V(t)$ and $I(t)$ are, respectively, voltage and current as a function of time,

- V_o, I_o, Z, and Y are complex numbers,
- Z is called impedance,
- Y is called admittance,
- Re indicates real part,
- is the angular frequency of the AC current,
- $j = \sqrt{-1}$ is the imaginary unit.

The impedance and admittance may be expressed as complex numbers that can be broken into real and imaginary parts:

$$Z = R + jX, \quad Y = G + jB$$

where R and G are resistance and conductance respectively, X is reactance, and B is susceptance. For ideal resistors, Z and Y reduce to R and G respectively, but for AC networks containing capacitors and inductors, X and B are nonzero.

$Z = 1/Y$ for AC circuits, just as $R = 1/G$ for DC circuits.

Frequency Dependence of Resistance

Another complication of AC circuits is that the resistance and conductance can be frequency-dependent. One reason, mentioned above is the skin effect (and the related proximity effect). Another reason is that the resistivity itself may depend on frequency.

Energy Dissipation and Joule Heating

Running current through a material with high resistance creates heat, in a phenomenon called Joule heating. In this picture, a cartridge heater, warmed by Joule heating, is glowing red hot.

Resistors (and other elements with resistance) oppose the flow of electric current; therefore, electrical energy is required to push current through the resistance. This electrical energy is dissipated, heating the resistor in the process. This is called *Joule heating* (after James Prescott Joule), also called *ohmic heating* or *resistive heating*.

The dissipation of electrical energy is often undesired, particularly in the case of transmission losses in power lines. High voltage transmission helps reduce the losses by reducing the current for a given power.

On the other hand, Joule heating is sometimes useful, for example in electric stoves and other electric heaters (also called *resistive heaters*). As another example, incandescent lamps rely on Joule heating: the filament is heated to such a high temperature that it glows "white hot" with thermal radiation (also called incandescence).

The formula for Joule heating is:

$$P = I^2 R$$

where P is the power (energy per unit time) converted from electrical energy to thermal energy, R is the resistance, and I is the current through the resistor.

Dependence of Resistance on other Conditions

Temperature Dependence

Near room temperature, the resistivity of metals typically increases as temperature is increased, while the resistivity of semiconductors typically decreases as temperature is increased. The resistivity of insulators and electrolytes may increase or decrease depending on the system.

As a consequence, the resistance of wires, resistors, and other components often change with temperature. This effect may be undesired, causing an electronic circuit to malfunction at extreme temperatures. In some cases, however, the effect is put to good use. When temperature-dependent resistance of a component is used purposefully, the component is called a resistance thermometer or thermistor. (A resistance thermometer is made of metal, usually platinum, while a thermistor is made of ceramic or polymer.)

Resistance thermometers and thermistors are generally used in two ways. First, they can be used as thermometers. By measuring the resistance, the temperature of the environment can be inferred. Second, they can be used in conjunction with Joule heating (also called self-heating). If a large current is running through the resistor, the resistor's temperature rises and therefore its resistance changes. Therefore, these components can be used in a circuit-protection role similar to fuses, or for feedback in circuits, or for many other purposes. In general, self-heating can turn a resistor into a nonlinear and hysteretic circuit element. If the temperature T does not vary too much, a linear approximation is typically used:

$$R(T) = R_0[1 + \alpha(T - T_0)]$$

where α is called the *temperature coefficient of resistance*, T_0 is a fixed reference temperature (usually room temperature), and R_0 is the resistance at temperature T_0.

The parameter α is an empirical parameter fitted from measurement data. Because the linear approximation is only an approximation, α is different for different reference temperatures. For this reason it is usual to specify the temperature that α was measured at with a suffix, such as α_{15}, and the relationship only holds in a range of temperatures around the reference.

The temperature coefficient α is typically $+3\times10^{-3}\ K^{-1}$ to $+6\times10^{-3}\ K^{-1}$ for metals near room temperature. It is usually negative for semiconductors and insulators, with highly variable magnitude.

Strain Dependence

Just as the resistance of a conductor depends upon temperature, the resistance of a conductor depends upon strain. By placing a conductor under tension (a form of stress that leads to strain in the form of stretching of the conductor), the length of the section of conductor under tension increases and its cross-sectional area decreases. Both these effects contribute to increasing the resistance of the strained section of conductor. Under compression (strain in the opposite direction), the resistance of the strained section of conductor decreases.

Light Illumination Dependence

Some resistors, particularly those made from semiconductors, exhibit *photoconductivity*, meaning that their resistance changes when light is shining on them. Therefore, they are called *photoresistors* (or *light dependent resistors*). These are a common type of light detector.

Superconductivity

Superconductors are materials that have exactly zero resistance and infinite conductance, because they can have $V=0$ and $I\neq0$. This also means there is no joule heating, or in other words no dissipation of electrical energy. Therefore, if superconductive wire is made into a closed loop, current flows around the loop forever. Superconductors require cooling to temperatures near 4K with liquid helium for most metallic superconductors like niobium–tin alloys, or cooling to temperatures near 77K with liquid nitrogen for the expensive, brittle and delicate ceramic high temperature superconductors. Nevertheless, there are many technological applications of superconductivity, including superconducting magnets.

Electrical resistivity, represented by the Greek letter ρ (rho), is a measure of how strongly a material opposes the flow of electric current. The lower the resistivity, the more readily the material permits the flow of electric charge.

Electrical conductivity is the reciprocal quantity of resistivity. Conductivity is a measure of how well a material conducts an electric current.

Electric conductivity may be represented by the Greek letter σ (sigma), κ (kappa), or γ (gamma).

This is a table of the electrical resistivity and electrical conductivity of several materials.

Material	ρ (Ω•m) at 20°C Resistivity	σ (S/m) at 20°C Conductivity
Silver	1.59×10^{-8}	6.30×10^{7}
Copper	1.68×10^{-8}	5.96×10^{7}
Annealed copper	1.72×10^{-8}	5.80×10^{7}
Gold	2.44×10^{-8}	4.10×10^{7}
Aluminum	2.82×10^{-8}	3.5×10^{7}
Calcium	3.36×10^{-8}	2.98×10^{7}
Tungsten	5.60×10^{-8}	1.79×10^{7}
Zinc	5.90×10^{-8}	1.69×10^{7}
Nickel	6.99×10^{-8}	1.43×10^{7}
Lithium	9.28×10^{-8}	1.08×10^{7}
Iron	1.0×10^{-7}	1.00×10^{7}
Platinum	1.06×10^{-7}	9.43×10^{6}
Tin	1.09×10^{-7}	9.17×10^{6}
Carbon steel	(10^{10})	1.43×10^{-7}
Lead	2.2×10^{-7}	4.55×10^{6}
Titanium	4.20×10^{-7}	2.38×10^{6}
Grain oriented electrical steel	4.60×10^{-7}	2.17×10^{6}
Manganin	4.82×10^{-7}	2.07×10^{6}
Constantan	4.9×10^{-7}	2.04×10^{6}
Stainless steel	6.9×10^{-7}	1.45×10^{6}
Mercury	9.8×10^{-7}	1.02×10^{6}
Nichrome	1.10×10^{-6}	9.09×10^{5}
GaAs	5×10^{-7} to 10×10^{-3}	5×10^{-8} to 10^{3}
Carbon (amorphous)	5×10^{-4} to 8×10^{-4}	1.25 to 2×10^{3}
Carbon (graphite)	2.5×10^{-6} to 5.0×10^{-6} //basal plane 3.0×10^{-3} ⊥basal plane	2 to 3×10^{5} //basal plane 3.3×10^{2} ⊥basal plane
Carbon (diamond)	1×10^{12}	$\sim10^{-13}$
Germanium	4.6×10^{-1}	2.17

Sea water	2×10^{-1}	4.8
Drinking water	2×10^{1} to 2×10^{3}	5×10^{-4} to 5×10^{-2}
Silicon	6.40×10^{2}	1.56×10^{-3}
Wood (damp)	1×10^{3} to 4	10^{-4} to 10^{-3}
Deionized water	1.8×10^{5}	5.5×10^{-6}
Glass	10×10^{10} to 10×10^{14}	10^{-11} to 10^{-15}
Hard rubber	1×10^{13}	10^{-14}
Wood (oven dry)	1×10^{14} to 16	10^{-16} to 10^{-14}
Sulfur	1×10^{15}	10^{-16}
Air	1.3×10^{16} to 3.3×10^{16}	3×10^{-15} to 8×10^{-15}
Paraffin wax	1×10^{17}	10^{-18}
Fused quartz	7.5×10^{17}	1.3×10^{-18}
PET	10×10^{20}	10^{-21}
Teflon	10×10^{22} to 10×10^{24}	10^{-25} to 10^{-23}

Factors that affect Electrical Conductivity

There are three main factors that affect the conductivity or resistivity of a material:

1. Cross-Sectional Area - If the cross-section of a material is large, it can allow more current to pass through it. Similarly, a thin cross-section restricts current flow.
2. Length of the Conductor - A short conductor allows current to flow at a higher rate than a long conductor. It's sort of like trying to move a lot of people through a hallway.
3. Temperature - Increasing temperature makes particles vibrate or move more. Increasing this movement (increasing temperature) decreases conductivity because the molecules are more likely to get in the way of current flow. At extremely low temperatures, some materials are superconductors.

Electrical Reactance

Reactance, in electricity, is the measure of the opposition that a circuit or a part of a circuit presents to electric current insofar as the current is varying or alternating. Steady electric currents flowing along conductors in one direction undergo opposition called electrical resistance, but no reactance. Reactance is present in addition to resistance when conductors carry alternating current. Reactance also occurs for short intervals when direct current is changing as it approaches or departs from steady flow, for example, when switches are closed or opened.

Reactance is of two types: inductive and capacitive. Inductive reactance is associated with the magnetic field that surrounds a wire or a coil carrying a current. An alternating current in such a conductor, or inductor, sets up an alternating magnetic field that in turn affects the current in, and the voltage (potential difference) across, that part of the circuit. An inductor essentially opposes changes in current, making changes in the current lag behind those in the voltage. The current builds up as the driving voltage is already decreasing, tends to continue on at maximum value when the voltage is reversing its direction, falls off to zero as the voltage is increasing to maximum in the opposite direction, and reverses itself and builds up in the same direction as the voltage even as the voltage is falling off again. Inductive reactance, a measure of this opposition to the current, is proportional to both the frequency f of the alternating current and a property of the inductor called inductance (symbolized by L and depending in turn on the inductor's dimensions, arrangement, and surrounding medium). Inductive reactance X_L equals 2π times the product of the frequency of the current and the inductance of the conductor, simply $X_L = 2\pi fL$. Inductive reactance is expressed in ohms. (The unit of frequency is hertz, and that of inductance is henry.)

Capacitive reactance, on the other hand, is associated with the changing electric field between two conducting surfaces (plates) separated from each other by an insulating medium. Such a set of conductors, a capacitor, essentially opposes changes in voltage, or potential difference, across its plates. A capacitor in a circuit retards current flow by causing the alternating voltage to lag behind the alternating current, a relationship in contrast to that caused by an inductor. The capacitive reactance, a measure of this opposition, is inversely proportional to the frequency f of the alternating current and to a property of the capacitor called capacitance (symbolized by C and depending on the capacitor's dimensions, arrangement, and insulating medium). The capacitive reactance X_c equals the reciprocal of the product of 2π, the frequency of the current, and the capacitance of that part of the circuit, simply $X_C = 1/(2\pi fC)$. Capacitive reactance has units of ohms. (The unit of capacitance is farad.)

Because inductive reactance X_L causes the voltage to lead the current and capacitive reactance X_C causes the voltage to lag behind the current, total reactance X is their difference—that is, $X = X_L - X_C$. The reciprocal of the reactance, $1/X$, is called the susceptance and is expressed in units of reciprocal ohm, called mho (*ohm* spelled backward).

Spitzer Resistivity

The Spitzer resistivity, or its opposite the Spitzer conductivity first formulated by Lyman Spitzer in 1950, is an equation showing the electrical resistance in a plasma decreases in proportion to the electron temperature as $T_e^{-3/2}$.

The Spitzer resistivity η is a classical model of electrical resistivity commonly used in plasma physics, based upon electron-ion collisions. It is given by:

$$\eta = \frac{\pi Z e^2 m^{1/2} \ln \Lambda}{(4\pi\varepsilon_0)^2 (k_B T)^{3/2}},$$

where Z is the ionization of nuclei, m is the electron mass, ε_0 is the electric permittivity of free space, Λ is the Coulomb logarithm, k_B is Boltzmann's constant, and T is the temperature in kelvins.

Sheet Resistance

Sheet resistance (also known as surface resistance or surface resistivity) is a common electrical property used to characterise thin films of conducting and semiconducting materials. It is a measure of the lateral resistance through a thin square of material, i.e. the resistance between opposite sides of a square. The key advantage of sheet resistance over other resistance measurements is that it is independent of the size of the square - enabling an easy comparison between different samples. Another advantage is that it can be measured directly using a four-point probe.

General Theory

Sheet resistance (R_S) is commonly defined as the resistivity (ρ) of a material divided by its thickness (t):

$$R_S = \frac{\rho}{t}$$

The units of this equation resolve to ohms (Ω); however, it actually represents the resistance between opposite sides of a square of a material (rather than bulk resistance). As such the units $\Omega/\square$ (ohms per square) are commonly used.

Examples of Applications

Sheet resistance is a critical property for any thin film of material in which electrical charges are intended to travel along (rather than pass through). For example, thin-film devices (such as perovskite solar cells or organic LEDs) require conducting electrodes which generally have thicknesses in the nanometre to micrometre range. Figure shows how charges move within an LED device. The electrodes must transport electrical charge laterally and need low sheet resistances to reduce losses during this process. This becomes even more important when attempting to scale up the size of these devices, as the electrical charges will have to travel further along the electrodes before they can be extracted.

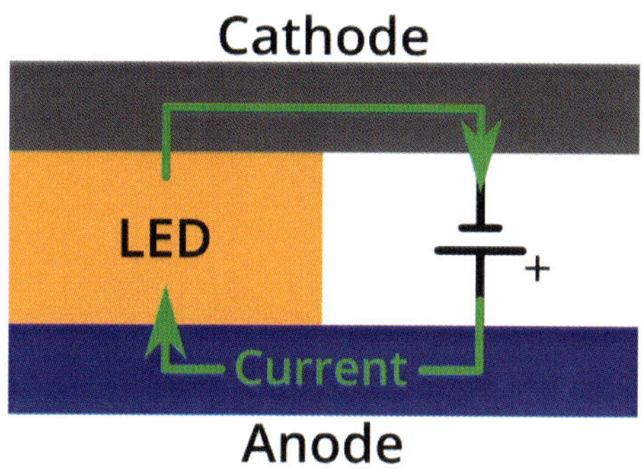

Figure A schematic diagram of a thin film LED, showing current flowing laterally through the electrodes to the active material. The sheet resistance of the electrodes will effect amount of current that reaches the LED, impacting its performance.

Furthermore, the resistivity and conductivity can be calculated if the sheet resistance and material thickness are known. This allows for the materials to be electrically characterised, purely by measuring their sheet resistance.

Measuring Sheet Resistance

The primary technique for measuring sheet resistance is the four-probe method (also known as the Kelvin technique), which is performed using a four-point probe. A four-point probe consists of four electrical probes in a line, with equal spacing between each of the probes as shown in Figure.

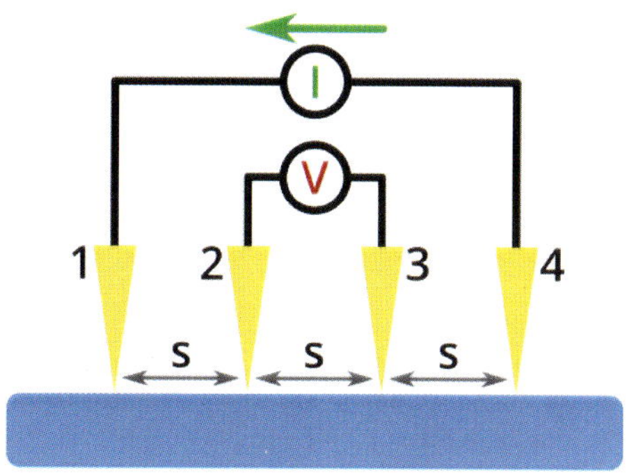

Figure A schematic diagram of a four-point probe. The four probes have equal spacing (s) and are shown in contact with a surface. A current (I) is injected through probe 1 and collected through probe 4, whilst the voltage is measured between probes 2 and 3.

It operates by applying a current (I) on the outer two probes and measuring the resultant voltage drop between the inner two probes. The sheet resistance can then be calculated using the equation below:

$$R_S = \frac{\pi}{\text{In}(2)} \frac{\Delta V}{I} = 4.53236 \frac{\Delta V}{I}$$

It should be noted that this equation is only valid if: i) the material being tested is no thicker than 40% of the spacing between the probes, and ii) the lateral size of the sample is sufficiently large. If this is not the case, then geometric correction factors are needed to account for the size, shape, and thickness of the sample. The value of this factor is dependent on the geometry being used.

Eliminating Contact Resistance

One of the primary advantages of using a four-point probe to perform electrical characterisation is the elimination of contact and wire resistances from the measurement. A diagram of the circuit resistances is shown in Figure below.

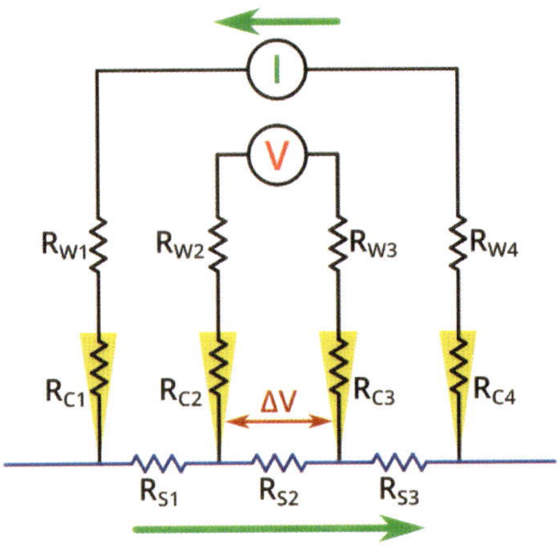

Figure Equivalent circuit diagram of a four-point probe, showing the wire resistances (R_W), contact resistances (R_C), and sample resistances (R_S). The green arrows represent current flow.

The applied current I enters and leaves the sample via the outer probes, and flows through the sample. Voltmeters typically have high electrical impedance to prevent them affecting the circuit being measured, so no current will flow through the inner two probes. Only the voltage is measured between the inner probes, meaning that the wire resistances (R_{W2} and R_{W3}) and the contact resistances (R_{C2} and R_{C3}) do not contribute to the measurement. Any measured decrease in voltage (ΔV) will therefore arise entirely from the sample resistance (R_{S2}). This simplifies the sheet resistance equation, so that only ΔV and the applied current are required to find the value of R_{S2} (i.e. the sheet resistance).

Geometric Correction Factors

Whilst the above equation for sheet resistance is independent of sample geometry, this only applies when the sample is significantly larger (typically having dimensions 40 times greater) than the spacing of the probes, and if the sample is thinner than 40% of the probe spacing. If this is not the case, the possible current paths between the probes are limited by the proximity to the edges of the sample, resulting in an overestimation of the sheet resistance. To account for this difference, a correction factor based upon the geometry of the sample is required.

Circular Samples

For a circular sample of diameter d, measured at the centre of the sample, the correction factor can be calculated using:

$$C = \frac{\text{In}(2)}{\text{In}(2) + \text{In}\left(d^2/_{s^2} + 3\right) - \text{In}\left(d^2/_{s^2} - 3\right)}$$

Where s is the distance between probes. For d >> **s** this equation tends to unity, enabling the use of the uncorrected equation.

Rectangular Samples

For a rectangular sample, the determination of the geometrical correction factor is slightly more complicated as there is no equation. Instead, a table of empirically-determined correction factors is used. The values in this table only apply when the probes make contact in the centre of the sample, and are aligned parallel to the sample's longest edge (l), as shown in Figure.

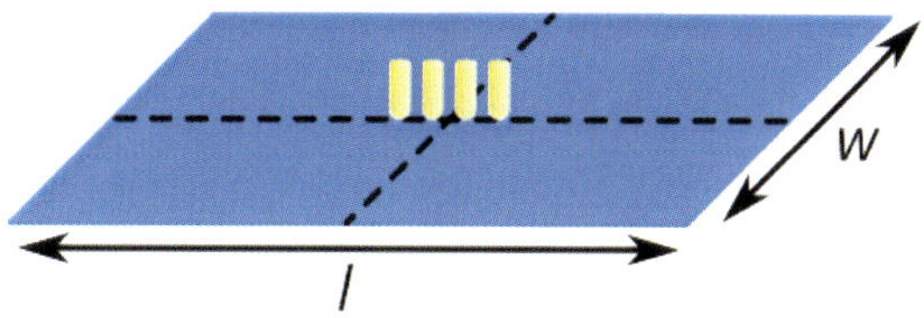

Figure Illustration of probe positioning and dimensions for a rectangular sample with l ≥ w.

As an example, suppose the rectangular sample shown in the figure above has a long edge of l = 20 mm and short edge of w = 10 mm, and the spacing of the probes being used is s = 2 mm. In this case, l / w = 2 and w / s = 5, so the table is searched for the correction factor which satisfies these two values, looking along the columns for l / w = 2 and the rows for w / s = 5, which is C = 0.7887. The measured sheet resistance is multiplied by this value to get the correct value for the sample.

w / s	l / w = 1	l / w = 2	l / w = 3	l / w = 4
1			0.2204	0.2205
1.25			0.2751	0.2751
1.5		0.3263	0.3286	0.3286
1.75		0.3794	0.3803	0.3803
2		0.4292	0.4297	0.4297
2.5		0.5192	0.5194	0.5194
3	0.5422	0.5957	0.5958	0.5958
4	0.6870	0.7115	0.7115	0.7115
5	0.7744	0.7887	0.7887	0.7887
7.5	0.8846	0.8905	0.8905	0.8905

10	0.9313	0.9345	0.9345	0.9345
15	0.9682	0.9696	0.9696	0.9696
20	0.9822	0.9830	0.9830	0.9830
40	0.9955	0.9957	0.9957	0.9957
∞	1	1	1	1

Obviously, not every sample will fall neatly into these categories. If this is the case, it is recommended that cubic spline interpolation is used to estimate the appropriate correction factor for the sample.

It is important to note that the correction factors for circular and rectangular samples detailed above only apply for measurements taken in the *centre of the sample*. If the measurement is not in the centre, different correction factors are needed.

Thick Samples

If the sample being tested is thicker than 40% of the probe spacing, an additional correction factor is required. The correction factor used is dependent upon the ratio of the sample thickness (t) to the probe spacing (**s**) and some of the possible values are listed in the table below:

t / s	**Correction Factor**
0.4	0.9995
0.5	0.9974
0.5555	0.9948
0.6250	0.9898
0.7143	0.9798
0.8333	0.9600
1.0	0.9214
1.1111	0.8907
1.25	0.8490
1.4286	0.7938
1.6666	0.7225
2.0	0.6336

As with the rectangular samples, if t / s does not equal one of the values given in the table, a cubic spline interpolation is recommended to estimate the appropriate correction factor for the sample.

Four-point Probe Equation Derivation

In order to determine how the sheet resistance of a thin film is measured using a four-point probe, a simplified scenario must first be evaluated. Imagine an arbitrarily sharp probe contacting and injecting current (through an applied voltage) into a semi-infinite volume (infinite in all directions except towards the probe) of a conductive material.

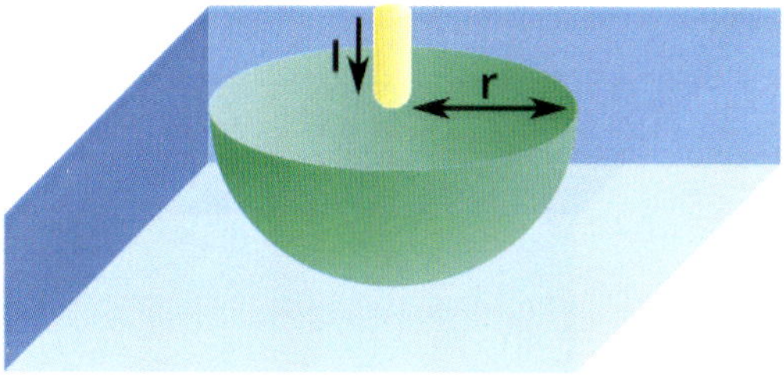

Figure A probe injecting current I into a semi-infinite volume of conductive material. The green hemisphere is a shell of the injected current with radius r.

The current travels outwards from the point of contact through concentric hemispherical shells of equipotential, each of which have current density (J) of:

$$J = \frac{I}{2\pi r^2}$$

Where r is the radial distance from the probe ($2\pi r^2$ being the surface area of the hemisphere). By applying Ohm's Law (E = ρJ) with the electric field across each shell equal to the voltage drop over the shell thickness, or - $\Delta \mathbf{V} \ / \ \Delta \mathbf{r}$ (this term is negative as voltage decreases with r), and with the thickness of the shell tending towards zero, the following equation is obtained:

$$\frac{dV}{dr} = -\rho\left(\frac{I}{2\pi r^2}\right)$$

This can be integrated between r and r' to obtain:

$$V - V' = \frac{\rho I}{2\pi}\left(\frac{1}{r} - \frac{1}{r'}\right)$$

By applying the boundary condition that V approaches zero as r approaches infinity, the equation simplifies to:

$$V = \frac{\rho I}{2\pi}$$

Now imagine that there are four arbitrarily sharp probes (labelled 1 to 4) in contact with the semi-infinite conducting material, which are in a line with equal spacing (s). They are set up so that current is injected through probe 1 and collected by probe 4 as shown in figure above. If equivalent boundary conditions are assumed for each probe, the voltage at any point is equal to the sum of the voltage due to each probe separately, i.e.:

$$V = \frac{\rho I}{2\pi}\left(\frac{1}{r_1} - \frac{1}{r_4}\right)$$

Where r_1 and r_4 are the radial distances from probe 1 and probe 4 respectively. Measurements of the voltage are then made between probes 2 and 3. Using the above equation, the voltage at probes 2 and 3 are:

$$V_2 = \frac{\rho I}{2\pi}\left(\frac{1}{s} - \frac{1}{2s}\right)$$

$$V_3 = \frac{\rho I}{2\pi}\left(\frac{1}{2s} - \frac{1}{s}\right)$$

Hence, the change in voltage (ΔV) between probes 2 and 3 is:

$$\Delta V = V_2 - V_3 = \frac{\rho I}{2\pi}\left(\frac{2}{s} - \frac{1}{s}\right) = \frac{\rho I}{2\pi s}$$

Therefore, the resistivity between the probes is:

$$\rho = 2\pi s\left(\frac{\Delta V}{I}\right)$$

This expression only applies in the case of a semi-infinite volume, and does not apply in the case of a thin film. However, a new expression can be derived using a similar analysis. As before, imagine the arbitrarily sharp probe contacting and injecting current into a thin film of material with thickness t.

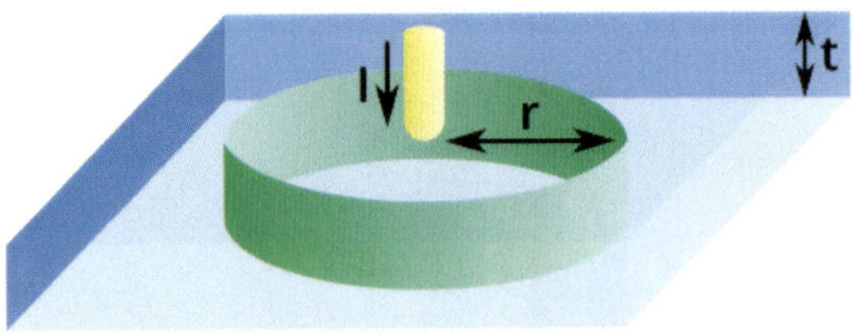

Figure A probe injecting current I into a thin film of conductive material with thickness t. The green cylinder is a shell of the injected current with radius r.

In this case, the current travels away from the probe (through the material) in short cylindrical shells of equipotential, each with a current density of:

$$J = \frac{I}{2\pi rt}$$

By applying the same conditions for the electric field as previously (Ohm's Law and shell thickness tending to zero), the electric field across each shell is:

$$\frac{dV}{dr} = -\rho\left(\frac{I}{2\pi rt}\right)$$

The resistivity has already been defined as the sheet resistance multiplied by the thickness of the materials, so this can be replaced in the above equation to give:

$$\frac{dV}{dr} = - \quad \left(\frac{I}{r}\right)$$

This can be integrated between r and r' to obtain:

$$V - V' = \frac{IR_S}{2\pi}\left(\mathrm{In}\frac{1}{r} - \mathrm{In}\frac{1}{r'}\right) = \frac{IR_S}{2\pi}\left(\mathrm{In}\, r' - \mathrm{In}\, r\right)$$

Unlike before, it cannot be assumed that the voltage tends to zero as r approaches infinity as the natural logarithm of infinity is not zero. However, this does not impact the analysis as the difference in voltage at different points (ΔV) is the value measured by the four-point probe.

Now imagine the four-probe system in contact with a thin film, with an additional condition that the thickness of the film (t) is negligible compared to the probe spacing. For current being injected by probe 1 and collected by probe 4, the equation becomes:

$$V = \frac{IR_S}{2\pi}\left(\mathrm{In}\, r_4 - \mathrm{In}\, r_1\right)$$

The voltages measured at probes 2 and 3 are therefore:

$$V_2 = \frac{IR_S}{2\pi}\left(\mathrm{In}\, 2s - \mathrm{In}\, s\right)$$

$$V_3 = \frac{IR_S}{2\pi}\left(\mathrm{In}\, s - \mathrm{In}\, 2s\right)$$

Hence, the change in voltage is:

$$\Delta V = V_2 - V_3 = \frac{IR_S}{2\pi}\left(2\mathrm{In}\, 2s - 2\mathrm{In}\, s\right) = \frac{IR_S}{\pi}\mathrm{In}\, 2$$

which can be rearranged to give:

$$R_S = \frac{\pi}{\mathrm{In}(2)}\frac{\Delta V}{I} = 4.53236\frac{\Delta V}{I}$$

Therefore, by measuring the change in voltage between the inner probes and the applied current between the outer probes, we can measure the sheet resistance of a sample.

Capacitance

Capacitance is a physical property of the capacitor (and thus is independent to the capacitor's particular charge or voltage at any given time). Capacitance is always positive. Capacitance is measured in farads (F), defined as 1 farad = 1 coulomb/volt. Like coulombs, farads are very large, so most real life capacitors are in the range of microfarads (10^{-6} F), nanofarads (10^{-9} F) or picofarads (10^{-12} F), though capacitors with values of several farads are available. Note that when a capacitor is charged, its two plates have opposite charges. When dealing with capacitors, Q is generally taken to represent the magnitude of the charge on either plate of the capacitor. The net charge of both plates of a capacitor is generally zero.

Formula for Capacitance

To measure capacitance of a material, if a voltage V is attached to the material, the capacitance C depends on this equation:

$$C = \frac{Q}{V}$$

Where Q is the charge in coulombs, C is the capacitance in farads and V is the voltage of the battery in volts.

The energy in joules stored in a capacitor when it is charged up to a voltage V is

$$E = \frac{1}{2} c V^2$$

Capacitance Calculation

We can use the formulas to calculate total capacitance in a circuit. Note that the formulas are basically the inverse of the formulas for resistors.

Calculating Capacitors in Parallel

To calculate the values of a string of capacitors in parallel, just add them up:

$$C_t = C_1 + C_2 + C_3 + \ldots..$$

In this formula, C_1, C_2, C_3 and so forth are the values of the capacitors; C_t is the total capacitance.

Calculating Two Capacitors in Series

To calculate the total capacitance of two capacitors in series is,

$$C_t = \frac{C_1 \times C_2}{C_1 + C_2}$$

In this formula, C_1 and C_2 are the values of the two capacitors; C_t is the total capacitance.

Calculating Three or More Capacitors in Series

To calculate the total capacitance of three capacitors is;

$$C_t = \frac{1}{\frac{1}{c_1} + \frac{1}{c_2} + \frac{1}{c_3}}$$

C_1, C_2, C_3 and so forth are the values of the capacitors. C_t is the total capacitance.

Self-capacitance

In electrical circuits, the term *capacitance* is usually a shorthand for the *mutual capacitance* between two adjacent conductors, such as the two plates of a capacitor. However, for an isolated conductor there also exists a property called *self-capacitance*, which is the amount of electric charge that must be added to an isolated conductor to raise its electric potential by one unit (i.e. one volt, in most measurement systems). The reference point for this potential is a theoretical hollow conducting sphere, of infinite radius, with the conductor centered inside this sphere.

Mathematically, the *self-capacitance* of a conductor is defined by

$$C = \frac{q}{V},$$

where

q is the charge held by the conductor,

$V = \frac{1}{4\pi\varepsilon_0}\int \frac{\sigma dS}{r}$ is the electric potential,

σ is the surface charge density.

dS is an infinitesimal element of area,

r is the length from dS to a fixed point M within the plate

Using this method, the self-capacitance of a conducting sphere of radius R is:

$$C = 4\pi\varepsilon_0 R$$

Example values of self-capacitance are:

- For the top "plate" of a van de Graaff generator, typically a sphere 20 cm in radius: 22.24 pF,
- The planet Earth: about 710 μF.

The inter-winding capacitance of a coil is sometimes called self-capacitance, but this is a different phenomenon. It is actually mutual capacitance between the individual turns of the coil and is a form of stray, or parasitic capacitance. This self-capacitance is an important consideration at high frequencies: It changes the impedance of the coil and gives rise to parallel resonance. In many applications this is an undesirable effect and sets an upper frequency limit for the correct operation of the circuit.

Mutual Capacitance

A common form is a parallel-plate capacitor, which consists of two conductive plates insulated from each other, usually sandwiching a dielectric material. In a parallel plate capacitor, capacitance is very nearly proportional to the surface area of the conductor plates and inversely proportional to the separation distance between the plates.

If the charges on the plates are $+q$ and $-q$, and V gives the voltage between the plates, then the capacitance C is given by

$$C = \frac{q}{V}.$$

which gives the voltage/current relationship

$$i(t) = C\frac{dv(t)}{dt}.$$

The energy stored in a capacitor is found by integrating the work W:

$$W_{\text{charging}} = \frac{1}{2}CV^2$$

Capacitance Matrix

The discussion above is limited to the case of two conducting plates, although of arbitrary size and shape. The definition $C = Q/V$ does not apply when there are more than two charged plates, or when the net charge on the two plates is non-zero. To handle this case, Maxwell introduced his *coefficients of potential.* If three (nearly ideal) conductors are given charges $Q_1, Q_2, Q_3,$, then the voltage at conductor 1 is given by

$$V_1 = P_{11}Q_1 + P_{12}Q_2 + P_{13}Q_3,$$

and similarly for the other voltages. Hermann von Helmholtz and Sir William Thomson showed that the coefficients of potential are symmetric, so that $P_{12} = P_{21}$, etc. Thus the system can be described by a collection of coefficients known as the *elastance matrix* or *reciprocal capacitance matrix*, which is defined as:

$$P_{ij} = \frac{\partial V_i}{\partial Q_j}$$

From this, the mutual capacitance C_m between two objects can be defined by solving for the total charge Q and using $C_m = Q/V$.

$$C_m = \frac{1}{(P_{11} + P_{22}) - (P_{12} + P_{21})}$$

Since no actual device holds perfectly equal and opposite charges on each of the two "plates", it is the mutual capacitance that is reported on capacitors.

The collection of coefficients $C_{ij} = \frac{\partial Q_i}{\partial V_j}$ is known as the *capacitance matrix*, and is the inverse of the elastance matrix.

Capacitors

The capacitance of the majority of capacitors used in electronic circuits is generally several orders of magnitude smaller than the farad. The most common subunits of capacitance in use today are the microfarad (μF), nanofarad (nF), picofarad (pF), and, in microcircuits, femtofarad (fF). However, specially made supercapacitors can be much larger (as much as hundreds of farads), and parasitic capacitive elements can be less than a femtofarad. In the past, alternate subunits were used in historical electronic books; "mfd" and "mf" for microfarad (μF); "mmfd", "mmf", " $\mu\mu F$ " for picofarad (pF); but are rarely used any more.

Capacitance can be calculated if the geometry of the conductors and the dielectric properties of the insulator between the conductors are known. A qualitative explanation for this can be given as follows.

Once a positive charge is put unto a conductor, this charge creates an electrical field, repelling any other positive charge to be moved onto the conductor; i.e., increasing the necessary voltage. But if nearby there is another conductor with a negative charge on it, the electrical field of the positive conductor repelling the second positive charge is weakened (the second positive charge also feels the attracting force of the negative charge). So due to the second conductor with a negative charge, it becomes easier to put a positive charge on the already positive charged first conductor, and vice versa; i.e., the necessary voltage is lowered.

As a quantitative example consider the capacitance of a capacitor constructed of two parallel plates both of area *A* separated by a distance *d*. If *d* is sufficiently small with respect to the smallest chord of *A*, there holds, to a high level of accuracy:

$$C = \varepsilon_r \varepsilon_0 \frac{A}{d}$$

where

C is the capacitance, in farads;

A is the area of overlap of the two plates, in square meters;

ε_r is the relative static permittivity (sometimes called the dielectric constant) of the material between the plates (for a vacuum, $\varepsilon_r = 1$);

ε_0 is the electric constant ($\varepsilon_0 \approx 8.854 \times 10^{-12}\ F \cdot m^{-1}$); and

d is the separation between the plates, in meters.

Capacitance is proportional to the area of overlap and inversely proportional to the separation between conducting sheets. The closer the sheets are to each other, the greater the capacitance. The equation is a good approximation if *d* is small compared to the other dimensions of the plates so that the electric field in the capacitor area is uniform, and the so-called *fringing field* around the periphery provides only a small contribution to the capacitance. In CGS units the equation has the form:

$$C = \varepsilon_r \frac{A}{4\pi d}$$

where *C* in this case has the units of length. Combining the SI equation for capacitance with the above equation for the energy stored in a capacitance, for a flat-plate capacitor the energy stored is:

$$W_{\text{stored}} = \frac{1}{2} C V^2 = \frac{1}{2} \varepsilon_r \varepsilon_0 \frac{A}{d} V^2$$

where *W* is the energy, in joules; *C* is the capacitance, in farads; and *V* is the voltage, in volts.

Stray Capacitance

Any two adjacent conductors can function as a capacitor, though the capacitance is small unless the conductors are close together for long distances or over a large area.

This (often unwanted) capacitance is called parasitic or "stray capacitance". Stray capacitance can allow signals to leak between otherwise isolated circuits (an effect called crosstalk), and it can be a limiting factor for proper functioning of circuits at high frequency.

Stray capacitance between the input and output in amplifier circuits can be troublesome because it can form a path for feedback, which can cause instability and parasitic oscillation in the amplifier. It is often convenient for analytical purposes to replace this capacitance with a combination of one input-to-ground capacitance and one output-to-ground capacitance; the original configuration — including the input-to-output capacitance — is often referred to as a pi-configuration. Miller's theorem can be used to effect this replacement: it states that, if the gain ratio of two nodes is $1/K$, then an impedance of Z connecting the two nodes can be replaced with a $Z/(1 - k)$ impedance between the first node and ground and a $KZ/(K - 1)$ impedance between the second node and ground. Since impedance varies inversely with capacitance, the internode capacitance, C, is replaced by a capacitance of KC from input to ground and a capacitance of $(K - 1)C/K$ from output to ground. When the input-to-output gain is very large, the equivalent input-to-ground impedance is very small while the output-to-ground impedance is essentially equal to the original (input-to-output) impedance.

Capacitance of Conductors with Simple Shapes

Calculating the capacitance of a system amounts to solving the Laplace equation $\nabla^2\varphi=0$ with a constant potential φ on the surface of the conductors. This is trivial in cases with high symmetry. There is no solution in terms of elementary functions in more complicated cases.

For two-dimensional situations analytic functions may be used to map different geometries to each other.

Capacitance of simple systems		
Type	Capacitance	Comment
Parallel-plate capacitor	$\varepsilon A / d$	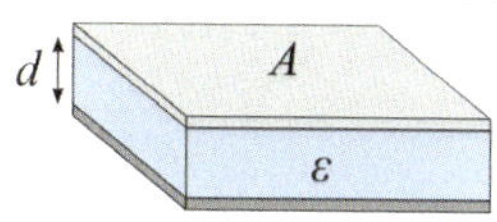 ε: Permittivity
Coaxial cable	$\dfrac{2\pi\varepsilon\ell}{\ln(R_2/R_1)}$	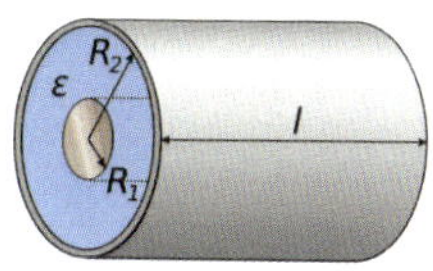 ε: Permittivity

Pair of parallel wires	$\frac{\pi\varepsilon\ell}{\operatorname{arcosh}\left(\frac{d}{2a}\right)}=\frac{\pi\varepsilon\ell}{\ln\left(\frac{d}{2a}+\sqrt{\frac{d^2}{4a^2}-1}\right)}$	
Wire parallel to wall	$\frac{2\pi\varepsilon\ell}{\operatorname{arcosh}\left(\frac{d}{a}\right)}=\frac{2\pi\varepsilon\ell}{\ln\left(\frac{d}{a}+\sqrt{\frac{d^2}{a^2}-1}\right)}$	a: Wire radius d: Distance, d > a ℓ: Wire length
Two parallel coplanar strips	$\varepsilon\ell\frac{K\left(\sqrt{1-k^2}\right)}{K(k)}$	d: Distance w_1, w_2: Strip width k_m: $d/(2w_m+d)$ k^2: k_1k_2 K: Elliptic integral l: Length
Concentric spheres	$\frac{4\pi\varepsilon}{\frac{1}{R_1}-\frac{1}{R_2}}$	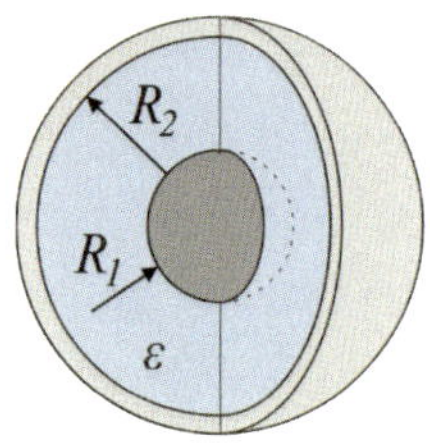 ε: Permittivity
Two spheres, equal radius	$2\pi\varepsilon a\sum_{n=1}^{\infty}\frac{\sinh\left(\ln\left(D+\sqrt{D^2-1}\right)\right)}{\sinh\left(n\ln\left(D+\sqrt{D^2-1}\right)\right)}$ $=2\pi\varepsilon a\left\{1+\frac{1}{2D}+\frac{1}{4D^2}+\frac{1}{8D^3}+\frac{1}{8D^4}+\frac{3}{32D^5}+O\left(\frac{1}{D^6}\right)\right\}$ $=2\pi\varepsilon a\left\{\ln 2+\gamma-\frac{1}{2}\ln(2D-2)+O(2D-2)\right\}$	a: Radius d: Distance, d > 2a D = d/2a, D > 1 γ: *Euler's constant*
Sphere in front of wall	$4\pi\varepsilon a\sum_{n=1}^{\infty}\frac{\sinh\left(\ln\left(D+\sqrt{D^2-1}\right)\right)}{\sinh\left(n\ln\left(D+\sqrt{D^2-1}\right)\right)}$	a: Radius d: Distance, d > a D = d/a
Sphere	$4\pi\varepsilon a$	a: Radius
Circular disc	$8\varepsilon a$	a: Radius
Prolate ellipsoid	$\frac{\varepsilon a}{\ln(2a/b)}$	half-axes a>b=c
Thin straight wire, finite length	$\frac{2\pi\varepsilon\ell}{\Lambda}\left\{1+\frac{1}{\Lambda}(1-\ln 2)+\frac{1}{\Lambda^2}\left[1+(1-\ln 2)^2-\frac{\pi^2}{12}\right]+O\left(\frac{1}{\Lambda^3}\right)\right\}$	a: Wire radius ℓ: Length Λ: ln(ℓ/a)

Energy Storage

The energy (measured in joules) stored in a capacitor is equal to the *work* required to push the charges into the capacitor, i.e. to charge it. Consider a capacitor of capacitance C, holding a charge $+q$ on one plate and $-q$ on the other. Moving a small element of charge dq from one plate to the other against the potential difference $V = q/C$ requires the work dW:

$$dW = \frac{q}{C}dq$$

where W is the work measured in joules, q is the charge measured in coulombs and C is the capacitance, measured in farads.

The energy stored in a capacitor is found by integrating this equation. Starting with an uncharged capacitance ($q = 0$) and moving charge from one plate to the other until the plates have charge $+Q$ and $-Q$ requires the work W:

$$W_{\text{charging}} = \int_0^Q \frac{q}{C}dq = \frac{1}{2}\frac{Q^2}{C} = \frac{1}{2}QV = \frac{1}{2}CV^2 = W_{\text{stored}}.$$

Nanoscale Systems

The capacitance of nanoscale dielectric capacitors such as quantum dots may differ from conventional formulations of larger capacitors. In particular, the electrostatic potential difference experienced by electrons in conventional capacitors is spatially well-defined and fixed by the shape and size of metallic electrodes in addition to the statistically large number of electrons present in conventional capacitors. In nanoscale capacitors, however, the electrostatic potentials experienced by electrons are determined by the number and locations of all electrons that contribute to the electronic properties of the device. In such devices, the number of electrons may be very small, however, the resulting spatial distribution of equipotential surfaces within the device are exceedingly complex.

Single-electron Devices

The capacitance of a connected, or "closed", single-electron device is twice the capacitance of an unconnected, or "open", single-electron device. This fact may be traced more fundamentally to the energy stored in the single-electron device whose "direct polarization" interaction energy may be equally divided into the interaction of the electron with the polarized charge on the device itself due to the presence of the electron and the amount of potential energy required to form the polarized charge on the device (the interaction of charges in the device's dielectric material with the potential due to the electron).

Few Electron Devices

The derivation of a "quantum capacitance" of a few-electron device involves the thermodynamic chemical potential of an N-particle system given by

$$\mu(N) = U(N) - U(N-1)$$

whose energy terms may be obtained as solutions of the Schrödinger equation. The definition of capacitance,

$$\frac{1}{C} \equiv \frac{\Delta V}{\Delta Q},$$

with the potential difference

$$\Delta V = \frac{\Delta\mu}{e} = \frac{\mu(N+\Delta N) - \mu(N)}{e}$$

may be applied to the device with the addition or removal of individual electrons,

$$\Delta N = 1 \text{ and } \Delta Q = e.$$

Then

$$C_Q(N) = \frac{e^2}{\mu(N+1) - \mu(N)} = \frac{e^2}{E(N)}$$

is the "quantum capacitance" of the device.

This expression of "quantum capacitance" may be written as

$$C_Q(N) = \frac{e^2}{U(N)}$$

which differs from the conventional expression where $W_{\text{stored}} = U$, the stored electrostatic potential energy,

$$C = \frac{Q^2}{2U}$$

by a factor of $1/2$ with $Q = Ne$.

However, within the framework of purely classical electrostatic interactions, the appearance of the factor of $1/2$ is the result of integration in the conventional formulation,

$$W_{\text{charging}} = U = \int_0^Q \frac{q}{C} \mathrm{d}q$$

which is appropriate since $dq = 0$ for systems involving either many electrons or metallic electrodes, but in few-electron systems, $dq \to \Delta Q = e$. The integral generally becomes a summation. One may trivially combine the expressions of capacitance and electrostatic interaction energy,

$$Q = CV \text{ and } U = QV,$$

respectively, to obtain,

$$C = Q\frac{1}{V} = Q\frac{Q}{U} = \frac{Q^2}{U}$$

which is similar to the quantum capacitance. A more rigorous derivation is reported in the literature. In particular, to circumvent the mathematical challenges of the spatially complex equipotential surfaces within the device, an *average* electrostatic potential experiences by *each* electron is utilized in the derivation.

The reason for apparent mathematical differences is understood more fundamentally as the potential energy, $U(N)$, of an isolated device (self-capacitance) is twice that stored in a "connected" device in the lower limit N=1. As N grows large, $U(N) \to U$. Thus, the general expression of capacitance is

$$C(N) = \frac{(Ne)^2}{U(N)}.$$

In nanoscale devices such as quantum dots, the "capacitor" is often an isolated, or partially isolated, component within the device. The primary differences between nanoscale capacitors and macroscopic (conventional) capacitors are the number of excess electrons (charge carriers, or electrons, that contribute to the device's electronic behavior) and the shape and size of metallic electrodes. In nanoscale devices, nanowires consisting of metal atoms typically do not exhibit the same conductive properties as their macroscopic, or bulk material, counterparts.

Capacitance in Electronic and Semiconductor Devices

In electronic and semiconductor devices, transient or frequency-dependent current between terminals contains both conduction and displacement components. Conduction current is related to moving charge carriers (electrons, holes, ions, etc.), while displacement current is caused by time-varying electric field. Carrier transport is affected by electric field and by a number of physical phenomena - such as carrier drift and diffusion, trapping, injection, contact-related effects, impact ionization, etc. As a result, device admittance is frequency-dependent, and a simple electrostatic formula for capacitance $C = q / V$, is not applicable. A more general definition of capacitance, encompassing electrostatic formula, is:

$$C = \frac{\mathrm{Im}(Y(\omega))}{\omega},$$

where $Y(\omega)$ is the device admittance, and ω is the angular frequency.

In general case, capacitance is a function of frequency. At high frequencies, capacitance approached a constant value, equal to "geometric" capacitance, determined by the terminals' geometry and dielectric content in the device. In particular, capacitance can be calculated by a Fourier transform of a transient current in response to a step-like voltage excitation:

$$C(w) = 1/(\Delta V)\int_0^{\infty} [i(t) - i(\infty)\cos(wt)\, dt.$$

Negative Capacitance in Semiconductor Devices

Usually, capacitance in semiconductor devices is positive. However, in some devices and under certain conditions (temperature, applied voltages, frequency, etc.), capacitance can become negative. Non-monotonic behavior of the transient current in response to a step-like excitation has been proposed as the mechanism of negative capacitance. Negative capacitance has been demonstrated and explored in many different types of semiconductor devices.

Inductance

An inductor is an electronic component consisting simply of a coil of wire. A constant electric current running through an inductor produces a magnetic field. If the current changes, so does the magnetic field. The unit for inductance is the henry (H), named after Joseph Henry, an American physicist who discovered inductance independently at about the same time as English physicist Michael Faraday. One henry is the amount of inductance that is required to induce one volt of electromotive force when the current is changing at one ampere per second.

Oersted's Law, named after Danish physicist Hans Christian Oersted, states that a constant electric current generates a magnetic field around the conductor. Faraday's Law of Induction states that a changing magnetic field induces a current in a conductor within that field. Finally, Lenz's law, named after Russian physicist Heinrich Lenz, states that this induced current is in the opposite direction of the change in current that produced the magnetic field. This phenomenon is called self-inductance.

What this means is, if you quickly reduce the current through the inductor, the changing magnetic field will induce a current that opposes the change, which tends to maintain the current at its previous level. Conversely, if you increase the current sharply, the induced current will be in the opposite direction of the increase, which again tends to maintain the current at a constant level. In other words, an inductor creates a kind of inertia in the current flow that resists rapid changes in much the same way that a massive body resists changes in its velocity.

One important application of inductors in active circuits is that they tend to block high-frequency signals while letting lower-frequency oscillations pass. Note that this is the opposite function of capacitors. Combining the two components in a circuit can selectively filter or generate oscillations of almost any desired frequency.

Source of Inductance

A current i flowing through a conductor generates a magnetic field around the conductor, which is described by Ampere's circuital law. The total magnetic flux through a circuit Φ is equal to the product of the magnetic field and the area of the surface spanning the current path. If the current varies, the magnetic flux Φ through the circuit changes. By Faraday's law of induction, any change in flux through a circuit induces an electromotive force (EMF) or voltage v in the circuit, proportional to the rate of change of flux.

$$v(t) = -\frac{d\Phi(t)}{dt}$$

The negative sign in the equation indicates that the induced voltage is in a direction which opposes the change in current that created it; this is called Lenz's law. The potential is therefore called a back EMF. If the current is increasing, the voltage is positive at the end of the conductor through which the current enters and negative at the end through which it leaves, tending to reduce the current. If the current is decreasing, the voltage is positive at the end through which the current leaves the conductor, tending to maintain the current. Self-inductance, usually just called inductance, L is the ratio between the induced voltage and the rate of change of the current.

$$v(t) = L\frac{di(t)}{dt}$$

Thus, inductance is a property of a conductor or circuit, due to its magnetic field, which tends to oppose changes in current through the circuit. The unit of inductance in the SI system is the henry (H), named after American scientist Joseph Henry, which is the amount of inductance which generates a voltage of one volt when the current is changing at a rate of one ampere per second.

All conductors have some inductance, which may have either desirable or detrimental effects in practical electrical devices. The inductance of a circuit depends on the geometry of the current path, and on the magnetic permeability of nearby materials; ferromagnetic materials with a higher permeability like iron near a conductor tend to increase the magnetic field and inductance. Any alteration to a circuit which increases the flux (total magnetic field) through the circuit produced by a given current increases the inductance, because inductance is also equal to the ratio of magnetic flux to current.

$$L = \frac{\Phi(i)}{i}$$

An inductor is an electrical component consisting of a conductor shaped to increase the magnetic flux, to add inductance to a circuit. Typically it consists of a wire wound into a coil or helix. A coiled wire has a higher inductance than a straight wire of the same length, because the magnetic field lines pass through the circuit multiple times, it has multiple flux linkages. The inductance is proportional to the square of the number of turns in the coil.

The inductance of a coil can be increased by placing a magnetic core of ferromagnetic material in the hole in the center. The magnetic field of the coil magnetizes the material of the core, aligning its magnetic domains, and the magnetic field of the core adds to that of the coil, increasing the flux through the coil. This is called a ferromagnetic core inductor. A magnetic core can increase the inductance of a coil by thousands of times.

If multiple electric circuits are located close to each other, the magnetic field of one can pass through the other; in this case the circuits are said to be *inductively coupled*. Due to Faraday's law of induction, a change in current in one circuit can cause a change in magnetic flux in another circuit and thus induce a voltage in another circuit. The concept of inductance can be generalized in this case by defining the mutual inductance $M_{i,j}$ of circuit i and circuit j as the ratio of voltage induced in circuit j to the rate of change of current in circuit i. This is the principle behind a *transformer*.

Self-inductance and Magnetic Energy

If the current through a conductor with inductance is increasing, a voltage $v(t)$ will be induced across the conductor with a polarity which opposes the current, as described above (this is in addition to any voltage drop caused by the conductor's resistance). The charges flowing through the circuit lose potential energy moving from the higher voltage to the lower voltage end. The energy from the external circuit required to overcome this "potential hill" is being stored in the increased magnetic field around the conductor. Therefore, any inductance with a current through it stores energy in its magnetic field. At any given time t the power $p(t)$ flowing into the magnetic field, which is equal to the rate of change of the stored energy U, is the product of the current $i(t)$ and voltage $v(t)$ across the conductor

$$p(t) = \frac{dU}{dt} = v(t)i(t)$$

From $v(t) = L\frac{di(t)}{dt}$ above

$$\frac{dU}{dt} = L(i)i\frac{di}{dt}$$

$$dU = L(i)idi$$

When there is no current, there is no magnetic field and the stored energy is zero. Neglecting resistive losses, the energy U (measured in joules, in SI) stored by an inductance with a current I through it is equal to the amount of work required to establish the current through the inductance from zero, and therefore the magnetic field. This is given by:

$$U = \int_0^I L(i)idi$$

If the inductance $L(I)$ is constant over the current range, the stored energy is

$$U = L\int_0^I idi$$

$$U = \frac{1}{2}LI^2$$

So therefore inductance is also proportional to how much energy is stored in the magnetic field for a given current. This energy is stored as long as the current remains constant. If the current decreases, the magnetic field will decrease, inducing a voltage in the conductor in the opposite direction, negative at the end through which current enters and positive at the end through which it leaves. This will return stored magnetic energy to the external circuit.

If ferromagnetic materials are located near the conductor, such as in an inductor with a magnetic core, the above equation is only valid for linear regions of the magnetic flux, at currents below the level at which the ferromagnetic material saturates, where the inductance is approximately constant. If the magnetic field in the inductor approaches the level at which the core saturates, the inductance will begin to change with current, and the integral equation must be used.

Inductive Reactance

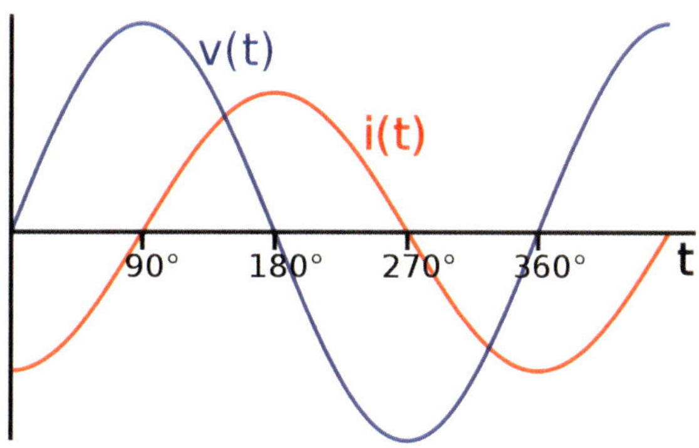

The voltage (v, *blue*) and current (*i, red*) waveforms in an ideal inductor to which an alternating current has been applied. The current lags the voltage by 90°

When a sinusoidal alternating current (AC) is passing through a linear inductance, the induced back-EMF will also be sinusoidal. If the current through the inductance is $i(t) = I_p \sin(2\pi ft)$, from (1) above the voltage across it will be

$$v(t) = L\frac{di}{dt} = L\frac{d}{dt}[I_p \sin(2\pi ft)] = 2\pi fLI_p \cos(2\pi ft) = 2\pi fLI_p \sin(2\pi ft + \frac{\pi}{2})$$

where I_p is the amplitude (peak value) of the sinusoidal current in amperes, f is the frequency of the alternating current in hertz, and L is the inductance. Thus the amplitude (peak value) of the voltage across the inductance will be

$$V_p = 2\pi fLI_p$$

Inductive reactance is the opposition of an inductor to an alternating current. It is defined analogously to electrical resistance in a resistor, as the ratio of the amplitude (peak value) of the alternating voltage to current in the component

$$X_L = \frac{V_p}{I_p} = 2\pi fL$$

Reactance has units of ohms. It can be seen that inductive reactance of an inductor increases proportionally with frequency f, so an inductor conducts less current for a given applied AC voltage as the frequency increases. Because the induced voltage is greatest when the current is increasing, the voltage and current waveforms are out of phase; the voltage peaks occur earlier in each cycle than the current peaks. The phase difference between the current and the induced voltage is $\phi = \pi / 2$ radians or 90 degrees, showing that in an ideal inductor *the current lags the voltage by 90°*.

Calculating Inductance

In the most general case, inductance can be calculated from Maxwell's equations. Many important cases can be solved using simplifications. Where high frequency currents are considered, with skin effect, the surface current densities and magnetic field may be obtained by solving the Laplace equation. Where the conductors are thin wires, self-inductance still depends on the wire radius and the distribution of the current in the wire. This current distribution is approximately constant (on the surface or in the volume of the wire) for a wire radius much smaller than other length scales.

Inductance of a Straight Single Wire

A straight single wire has some inductance, which in our ordinary experience is intangible because it is negligibly small so it can't readily be measured at low frequencies, and its effect is not detectable. A long straight wire like an electric transmission line has substantial inductance that reduces its capacity, and there is no problem at all measuring it. As a practical matter, longer wires have more inductance, and thicker wires have less, analogous to their electrical resistance, though the relationships aren't linear nor are they the same relationships as those quantities bear to resistance. As an essential component of coils and circuits, understanding what the inductance of a wire is, is essential. Yet, there is no simple answer.

There is no unambiguous definition of the inductance of a straight wire. If we consider the wire in isolation we ignore the question of how the current gets to the wire. That current will affect the flux which is developed in the vicinity of the wire. But this flux is a part of the definition. A consequence of Maxwell's equations is that we cannot define the inductance of only a portion of a circuit, we can only define the inductance of a whole circuit, which includes how the current gets to the wire and how it returns to the source. The magnetic flux incident to the whole circuit determines the inductance of the circuit and of any part of it. The magnetic flux is an indivisible entity, yet we wish to consider only a part of it, the part incident to the wire, between whatever we define to be the "ends" of the wire.

The total low frequency inductance (internal plus external) of a straight wire is:

$$L_{dc} = 2l \cdot [\ln(2l/r) - 0.75]$$

where

- L_{dc} is the "low-frequency" or DC inductance in nanohenries (nH or 10^{-9}H),
- l is the length of the wire in cm
- r is the radius of the wire in cm.

This result is based on the assumption that the radius r is much less than the length l, which is commonly true.

For sufficiently high frequencies skin effects cause the internal inductance to go to zero and the inductance becomes:

$$L_{ac} = 2l \cdot [\ln(2l/r) - 1.00]$$

These inductances are often referred to as "partial inductances" to indicate that they must be used with care.

In an everyday notion, one conductor of a 100m 18gauge lamp cord, stretched out straight, would have inductance of about 0.24mH.

Mutual Inductance of Two Parallel Straight Wires

There are two cases to consider: current travels in the same direction in each wire, and current travels in opposing directions in the wires. Currents in the wires need not be equal, though they often are, as in the case of a complete circuit, where one wire is the source and the other the return.

Mutual Inductance of Two Wire Loops

This is the generalized case of the paradigmatic 2-loop cylindrical coil carrying a uniform low frequency current; the loops are independent closed circuits that can have different lengths, any orientation in space, and carry different currents. None-the-less, the error terms, which are not included in the integral will only be small if the geometries of the loops are mostly smooth and convex: they do not have too many kinks, sharp corners, coils, crossovers, parallel segments, concave cavities or other topological "close" deformations. A necessary predicate for the reduction of the 3-dimensional manifold integration formula to a double curve integral is that the current paths be filamentary circuits, i.e. thin wires where the radius of the wire is negligible compared to its length.

The mutual inductance by a filamentary circuit m on a filamentary circuit n is given by the double integral *Neumann formula*

$$L_{m,n} = \frac{\mu_0}{4\pi} \oint_{C_m} \oint_{C_n} \frac{d\mathbf{x}_m \cdot d\mathbf{x}_n}{|\mathbf{x}_m - \mathbf{x}_n|}$$

where

- C_m and C_n are the curves spanned by the wires.
- μ_0 is the permeability of free space ($4\pi \times 10^{-7}$ H/m)
- $d\mathbf{x}_m$ is a small increment of the wire in circuit C_m
- $\mathbf{x}_m$ is the position of $d\mathbf{x}_m$ in space
- $d\mathbf{x}_n$ is a small increment of the wire in circuit C_n
- $\mathbf{x}_n$ is the position of $d\mathbf{x}_n$ in space

Derivation

$$M_{ij} \overset{\text{def}}{=} \frac{\Phi_i}{I_j}$$

where

- Φ_i is the magnetic flux through the ith surface by the electrical circuit outlined by C_j
- I_j is the current through the jth wire, this current creates the magnetic flux through the ith surface.

$$\Phi_i = \int_{S_i} \mathbf{B} \cdot \mathbf{da} = \int_{S_i} (\nabla \times \mathbf{A}) \cdot \mathbf{da} = \oint_{C_i} \mathbf{A} \cdot \mathbf{ds} = \oint_{c_i} \left(\sum_j \frac{\mu_0 I_j}{4\pi} \oint_{c_j} \frac{\mathbf{ds}_j}{|\mathbf{R}|} \right) \cdot \mathbf{ds}_i$$

where

C_i is the enclosing curve of S_i; and S_i is undefined

B is the magnetic field vector.

A is the vector potential.

Stokes' theorem has been used.

Self-inductance of a Wire Loop

Formally, the self-inductance of a wire loop would be given by the above equation with $m=n$. However, here $1/|\mathbf{x}-\mathbf{x}'|$ becomes infinite, leading to a logarithmically divergent integral. This necessitates taking the finite wire radius a and the distribution of the current in the wire into account. There remain the contribution from the integral over all points and a correction term,

$$L=\frac{\mu_0}{4\pi}\left[\oint_C\oint_{C'}\frac{d\mathbf{x}\cdot d\mathbf{x}'}{|\mathbf{x}-\mathbf{x}'|}\right]+\frac{\mu_0}{4\pi}lY+O \quad for\ |\mathbf{s}-\mathbf{s}'|>a/2$$

where

- $\mathbf{s}$ and $\mathbf{s}'$ are distances along the curves C and C' respectively
- a is the radius of the wire
- l is the length of the wire
- Y is a constant that depends on the distribution of the current in the wire: $Y=0$ when the current flows on the surface of the wire (total skin effect), $Y=1/2$ when the current is homogeneous over the cross-section of the wire.
- O is an error term $O(\mu_0 a)$ when the loop has sharp corners, and $O\left(\mu_0 a^2/l\right)$ when it is a smooth curve. These are small when the wire is long compared to its radius.

Inductance of a Solenoid

A solenoid is a long, thin coil; i.e., a coil whose length is much greater than its diameter. Under these conditions, and without any magnetic material used, the magnetic flux density B within the coil is practically constant and is given by

$$B=\frac{\mu_0 Ni}{l}$$

where μ_0 is the magnetic constant, N the number of turns, i the current and l the length of the coil. Ignoring end effects, the total magnetic flux through the coil is obtained by multiplying the flux density B by the cross-section area A:

$$\Phi = \frac{\mu_0 NiA}{l},$$

When this is combined with the definition of inductance $L = \frac{N\Phi}{i}$, it follows that the inductance of a solenoid is given by:

$$L = \frac{\mu_0 N^2 A}{l}.$$

Therefore, for air-core coils, inductance is a function of coil geometry and number of turns, and is independent of current.

Inductance of a Coaxial Cable

Let the inner conductor have radius r_i and permeability μ_i, let the dielectric between the inner and outer conductor have permeability μ_d, and let the outer conductor have inner radius r_{o1}, outer radius r_{o2}, and permeability μ_o. However, for a typical coaxial line application, we are interested in passing (non-DC) signals at frequencies for which the resistive skin effect cannot be neglected. In most cases, the inner and outer conductor terms are negligible, in which case one may approximate

$$L' = \frac{dL}{dl} \approx \frac{\mu_d}{2\pi} \ln \frac{r_{o1}}{r_i}$$

Inductance of Multilayer Coils

Most practical air-core inductors are multilayer cylindrical coils with square cross-sections to minimize average distance between turns (circular cross -sections would be better but harder to form).

Magnetic Cores

Many inductors include a magnetic core at the center of or partly surrounding the winding. Over a large enough range these exhibit a nonlinear permeability with effects such as magnetic saturation. Saturation makes the resulting inductance a function of the applied current.

The secant or large-signal inductance is used in flux calculations. It is defined as:

$$L_s(i) \stackrel{\text{def}}{=} \frac{N\Phi}{i} = \frac{\Lambda}{i}$$

The differential or small-signal inductance, on the other hand, is used in calculating voltage. It is defined as:

$$L_d(i) \stackrel{\text{def}}{=} \frac{d(N\Phi)}{di} = \frac{d\Lambda}{di}$$

The circuit voltage for a nonlinear inductor is obtained via the differential inductance as shown by Faraday's Law and the chain rule of calculus.

$$v(t) = \frac{d\Lambda}{dt} = \frac{d\Lambda}{di}\frac{di}{dt} = L_d(i)\frac{di}{dt}$$

Similar definitions may be derived for nonlinear mutual inductance.

Mutual Inductance

Derivation of Mutual Inductance

The inductance equations above are a consequence of Maxwell's equations. For the important case of electrical circuits consisting of thin wires, the derivation is straightforward.

In a system of K wire loops, each with one or several wire turns, the flux linkage of loop m, λ_m is given by

$$\lambda_m = N_m \Phi_m = \sum_{n=1}^{K} L_{m,n} i_n .$$

Here N_m denotes the number of turns in loop m; Φ_m, the magnetic flux through loop m; *and* $L_{m,n}$ are some constants. This equation follows from Ampere's law – magnetic fields and fluxes are linear functions of the currents. By Faraday's law of induction, we have

$$v_m = \frac{d\lambda_m}{dt} = N_m \frac{d\Phi_m}{dt} = \sum_{n=1}^{K} L_{m,n} \frac{di_n}{dt},$$

where v_m denotes the voltage induced in circuit m. This agrees with the definition of inductance above if the coefficients $L_{m,n}$ are identified with the coefficients of inductance. Because the total currents $N_n i_n$ contribute to Φ_m it also follows that $L_{m,n}$ is proportional to the product of turns $N_m N_n$.

Mutual Inductance and Magnetic Field Energy

Multiplying the equation for v_m above with $i_m dt$ and summing over m gives the energy transferred to the system in the time interval dt,

$$\sum_{m}^{K} i_m v_m dt = \sum_{m,n=1}^{K} i_m L_{m,n} di_n \overset{!}{=} \sum_{n=1}^{K} \frac{\partial W(i)}{\partial i_n} di_n .$$

This must agree with the change of the magnetic field energy, W, caused by the currents. The integrability condition

$$\frac{\partial^2 W}{\partial i_m \partial i_n} = \frac{\partial^2 W}{\partial i_n \partial i_m}$$

requires $L_{m,n} = L_{n,m}$. The inductance matrix, $L_{m,n}$, thus is symmetric. The integral of the energy transfer is the magnetic field energy as a function of the currents,

$$W(i) = \frac{1}{2}\sum_{m,n=1}^{K} i_m L_{m,n} i_n.$$

This equation also is a direct consequence of the linearity of Maxwell's equations. It is helpful to associate changing electric currents with a build-up or decrease of magnetic field energy. The corresponding energy transfer requires or generates a voltage. A mechanical analogy in the $K = 1$ case with magnetic field energy $(1/2)Li^2$ is a body with mass M, velocity u and kinetic energy $(1/2)Mu^2$. The rate of change of velocity (current) multiplied with mass (inductance) requires or generates a force (an electrical voltage).

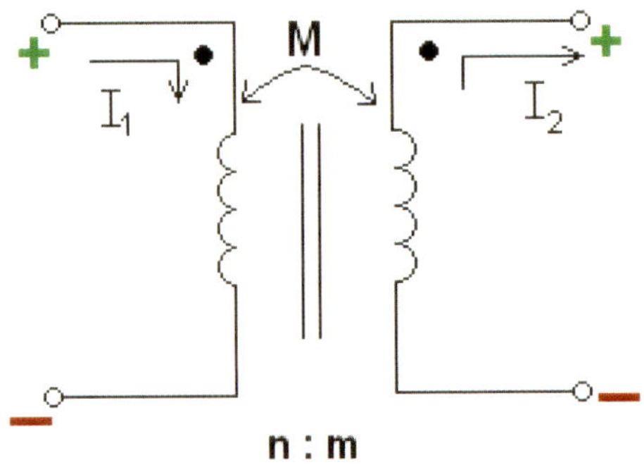

Circuit diagram of two mutually coupled inductors. The two vertical lines between the windings indicate that the transformer has a ferromagnetic core. "n:m" shows the ratio between the number of windings of the left inductor to windings of the right inductor. This picture also shows the dot convention.

Mutual inductance occurs when the change in current in one inductor induces a voltage in another nearby inductor. It is important as the mechanism by which transformers work, but it can also cause unwanted coupling between conductors in a circuit.

The mutual inductance, M, is also a measure of the coupling between two inductors. The mutual inductance by circuit i on circuit j is given by the double integral *Neumann formula*.

The mutual inductance also has the relationship:

$$M_{21} = N_1 N_2 P_{21}$$

where

M_{21} is the mutual inductance, and the subscript specifies the relationship of the voltage induced in coil 2 due to the current in coil 1.

N_1 is the number of turns in coil 1,

N_2 is the number of turns in coil 2,

P_{21} is the permeance of the space occupied by the flux.

Once the mutual inductance, *M*, is determined, it can be used to predict the behavior of a circuit:

$$v_1 = L_1 \frac{di_1}{dt} - M \frac{di_2}{dt}$$

where

v_1 is the voltage across the inductor of interest,

L_1 is the inductance of the inductor of interest,

di_1/dt is the derivative, with respect to time, of the current through the inductor of interest,

di_2/dt is the derivative, with respect to time, of the current through the inductor that is coupled to the first inductor, and

M is the mutual inductance.

The minus sign arises because of the sense the current i_2 has been defined in the diagram. With both currents defined going into the dots the sign of M will be positive (the equation would read with a plus sign instead).

Coupling Coefficient

The coupling coefficient is the ratio of the open-circuit actual voltage ratio to the ratio that would obtain if all the flux coupled from one circuit to the other. The coupling coefficient is related to mutual inductance and self inductances in the following way. From the two simultaneous equations expressed in the 2-port matrix the open-circuit voltage ratio is found to be:

$$\frac{V_2}{V_1}(\text{open circuit}) = \frac{M}{L_1}$$

while the ratio if all the flux is coupled is the ratio of the turns, hence the ratio of the square root of the inductances

$$\frac{V_2}{V_1}(\text{max coupled}) = \sqrt{\frac{L_2}{L_1}}$$

thus,

$$M = k\sqrt{L_1 L_2}$$

Where

k is the *coupling coefficient,*

L_1 is the inductance of the first coil, and

L_2 is the inductance of the second coil.

The coupling coefficient is a convenient way to specify the relationship between a certain orientation of inductors with arbitrary inductance. Most authors define the range as $0 \leq k < 1$, but some define it as $-1 < k < 1$. Allowing negative values of k captures phase inversions of the coil connections and the direction of the windings.

Equivalent Circuits

T-circuit

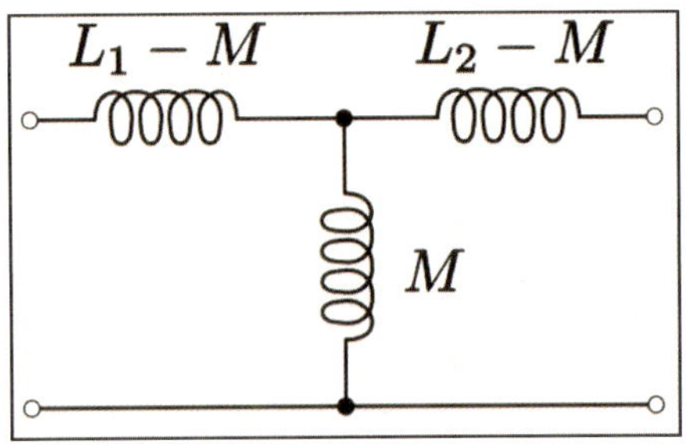

T equivalent circuit of mutually coupled inductors

Mutually coupled inductors can equivalently be represented by a T-circuit of inductors as shown. If the coupling is strong and the inductors are of unequal values then the series inductor on the step-down side may take on a negative value.

This can be analyzed as a two port network. With the output terminated with some arbitrary impedance, Z, the voltage gain, A_v, is given by,

$$A_v = \frac{sMZ}{s^2 L_1 L_2 - s^2 M^2 + sL_1 Z} = \frac{k}{s\left(1-k^2\right)\frac{\sqrt{L_1 L_2}}{Z} + \sqrt{\frac{L_1}{L_2}}}$$

where k is the coupling constant and s is the complex frequency variable, as above. For tightly coupled inductors where $k = 1$ this reduces to:

$$A_v = \sqrt{\frac{L_2}{L_1}}$$

which is independent of the load impedance. If the inductors are wound on the same core and with the same geometry, then this expression is equal to the turns ratio of the two inductors because inductance is proportional to the square of turns ratio.

The input impedance of the network is given by,

$$Z_{\text{in}} = \frac{s^2 L_1 L_2 - s^2 M^2 + sL_1 Z}{sL_2 + Z} = \frac{L_1}{L_2} Z\left(\frac{1}{1+\left(\frac{Z}{sL_2}\right)}\right)\left(1+\frac{\left(1-k^2\right)}{\left(\frac{Z}{sL_2}\right)}\right)$$

For $k = 1$ this reduces to:

$$Z_{\text{in}} = \frac{sL_1 Z}{sL_2 + Z} = \frac{L_1}{L_2} Z\left(\frac{1}{1+\left(\frac{Z}{sL_2}\right)}\right)$$

Thus, the current gain, A_i is *not* independent of load unless the further condition:

$$|sL_2| \gg |Z|$$

is met, in which case,

$$Z_{\text{in}} \approx \frac{L_1}{L_2} Z$$

and

$$A_i \approx \sqrt{\frac{L_1}{L_2}} = \frac{1}{A_v}$$

π-circuit

Alternatively, coupled inductors can be modelled using a π equivalent circuit as shown for two inductors.

Resonant Transformer

When a capacitor is connected across one winding of a transformer, making the winding a tuned circuit (resonant circuit) it is called a single-tuned transformer. When a capacitor is connected across each winding, it is called a double tuned transformer. These *resonant transformers* can store oscillating electrical energy similar to a resonant circuit and thus function as a bandpass filter, allowing frequencies near their resonant frequency to pass from the primary to secondary winding, but blocking other frequencies. The amount of mutual inductance between the two windings, together with the Q factor of the circuit, determine the shape of the frequency response curve. The advantage of the double tuned transformer is that it can have a narrower bandwidth than a simple tuned circuit. The coupling of double-tuned circuits is described as loose-, critical-, or over-coupled depending on the value of the coupling coefficient k. When two tuned circuits are loosely coupled through mutual inductance, the bandwidth will be narrow. As the amount of mutual inductance increases, the bandwidth continues to grow. When the mutual inductance is increased beyond the critical coupling, the peak in the frequency response curve splits into two peaks, and as the coupling is increased the two peaks move further apart. This is known as over coupling.

Ideal Transformers

When $k = 1$, the inductor is referred to as being closely coupled. If in addition, the self-inductances go to infinity, the inductor becomes an ideal transformer. In this case the voltages, currents, and number of turns can be related in the following way:

$$V_s = \frac{N_s}{N_p} V_p \text{ a}$$

where

V_s is the voltage across the secondary inductor,

V_p is the voltage across the primary inductor (the one connected to a power source),

N_s is the number of turns in the secondary inductor, and

N_p is the number of turns in the primary inductor.

Conversely the current:

$$I_s = \frac{N_p}{N_s} I_p$$

where

I_s is the current through the secondary inductor,

I_p is the current through the primary inductor (the one connected to a power source),

N_s is the number of turns in the secondary inductor, and

N_p is the number of turns in the primary inductor.

The power through one inductor is the same as the power through the other. These equations neglect any forcing by current sources or voltage sources.

Self-inductance of Thin Wire Shapes

The table below lists formulas for the self-inductance of various simple shapes made of thin cylindrical conductors (wires). In general these are only accurate if the wire radius a is much smaller than the dimensions of the shape, and if no ferromagnetic materials are nearby (no magnetic core).

Self-inductance of thin wire shapes		
Type	**Inductance**	**Comment**
Single layer solenoid	The well-known Wheeler's approximation formula for current-sheet model air-core coil: $L=\frac{N^2r^2}{9r+10l}$ (English) $L=\frac{N^2D^2}{45D+100l}(cgs)$ (cgs) his formula gives an error no more than 1% when $l/r>0.8$ or $l/D>0.4$.	• L: inductance in µH (10^{-6}H) • N: number of turns • r: radius in inches • D: diameter in cm • l: length in inches/cm
Coaxial cable (HF)	$\frac{\mu_0 l}{2\pi}\ln\left(\frac{b}{a}\right)$	b : Outer radius a : Inner radius l : Length
Circular loop	$\mu_0 r\left[\ln\left(\frac{8r}{a}\right)-2+\frac{Y}{4}+O\left(\frac{a^2}{r^2}\right)\right]$	r : Loop radius a : Wire radius
Rectangle made of round wire	$\frac{\mu_0}{\pi}\left[b\ln\left(\frac{2b}{a}\right)+d\ln\left(\frac{2d}{a}\right)+2\sqrt{b^2+d^2}\right.$ $\left.-b\sinh^{-1}\left(\frac{b}{d}\right)-d\sinh^{-1}\left(\frac{d}{b}\right)\quad-\left(2-\frac{Y}{4}\right)(b+d)\right]$	b,d: Border length $d\gg a, b\gg a$ a: Wire radius
Pair of parallel wires	$\frac{\mu_0 l}{\pi}\left[\ln\left(\frac{d}{a}\right)+\frac{Y}{4}\right]$	a: Wire radius d: Distance, $d\geq 2a$ l : Length of pair
Pair of parallel wires (HF)	$\frac{\mu_0 l}{\pi}\cosh^{-1}\left(\frac{d}{2a}\right)=\frac{\mu_0 l}{\pi}\ln\left(\frac{d}{2a}+\sqrt{\frac{d^2}{4a^2}-1}\right)$	a: Wire radius d: Distance, $d\geq 2a$ l : Length of pair

- The symbol μ_0 denotes the magnetic constant ($4\pi\times10^{-7}$ H/m) in SI units.
- Y is a constant between 0 and 1 that depends on the distribution of the current in the wire: $Y=0$ when the current flows on the surface of the wire (total skin effect), $Y=1$ when the current is homogeneous over the cross-section of the wire (direct current).

- $O(x)$ is represents small term(s) that have been dropped from the formula, to make it simpler. Read the symbol " $+O(x)$ " as "*plus small corrections on the order of*" x.

Impedance

Impedance (symbol Z) is a measure of the overall opposition of a circuit to current, in other words: how much the circuit impedes the flow of charge. It is like resistance, but it also takes into account the effects of capacitance and inductance. Impedance is measured in ohms (Ω).

Impedance is more complex than resistance because the effects of capacitance and inductance vary with the frequency of the current passing through the circuit and this means impedance varies with frequency. The effect of resistance is constant regardless of frequency.

Impedance, $Z = \frac{V}{I}$

Resistance, $R = \frac{V}{I}$

V = voltage in volts (V) I = current in amps (A)

Z = impedance in ohms (Ω)

R = resistance in ohms (Ω)

Impedance can be split into two parts:

- Resistance R (the part which is constant regardless of frequency)
- Reactance X (the part which varies with frequency due to capacitance and inductance)

Impedance in Simple Circuits

The term 'impedance' is often used (quite correctly) for simple circuits which have no capacitance or inductance, for example to refer to their 'input impedance' or 'output impedance'. This may seem confusing at first but for these simple circuits you can assume it is just another word for resistance.

The capacitance and inductance cause a phase shift between the current and voltage which means that the resistance and reactance cannot be simply added up to give impedance. Instead they must be added as vectors with reactance at right angles to resistance as shown in the diagram.

Four electrical quantities determine the impedance (Z) of a circuit: resistance (R), capacitance (C), inductance (L) and frequency (f).

The following topic on reactance explains how capacitance, inductance and frequency affect impedance.

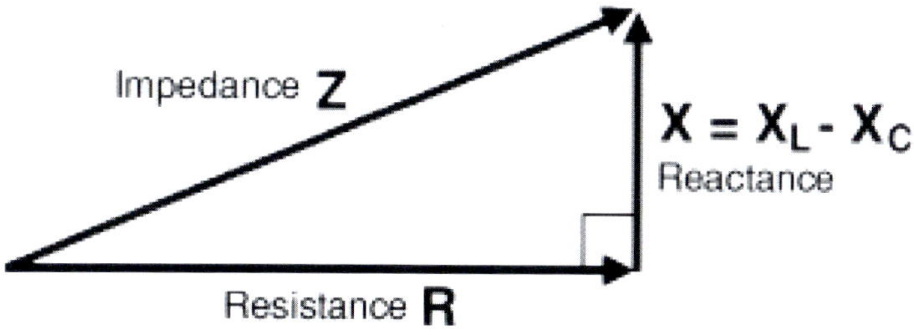

$$\text{Impedance,} \quad Z = \sqrt{R^2 + X^2}$$

Phase Shift

Phase shift means that the current and voltage are out of step with each other. Think of charging a capacitor. When the voltage across the capacitor is zero, the current is at a maximum; when the capacitor has charged and the voltage is at a maximum, the current is at a minimum. The charging and discharging occur continually with AC and the current reaches its maximum shortly before the voltage reaches its maximum: so we say the current leads the voltage.

Input Impedance Z_{IN}

Input impedance (Z_{IN}) is the impedance 'seen' by anything connected to the input of a circuit or device (such as an amplifer). It is the combined effect of all the resistance, capacitance and inductance connected to the input inside the circuit or device.

It is normal to use the term 'input impedance' even for simple cases where there is only resistance and the term 'input resistance' could be used instead. In fact it is usually reasonable to assume that an input impedance is just resistance providing the input signal has a low frequency (less than 1kHz).

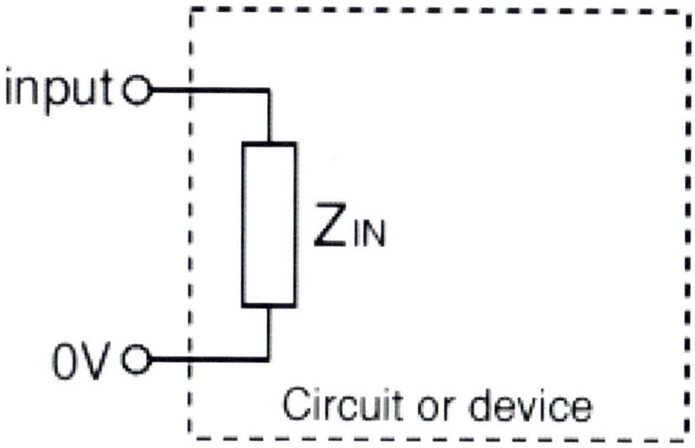

The effects of capacitance and inductance vary with frequency, so if these are present the input impedance will vary with frequency. The effects of capacitance and inductance are generally most significant at high frequencies.

Usually, input impedances should be high, at least ten times the output impedance of the circuit (or component) supplying a signal to the input. This ensures that the input will not 'overload' the source of the signal and reduce the strength (voltage) of the signal by a substantial amount.

Output Impedance Z_{OUT}

The output of any circuit or device is equivalent to an output impedance (Z_{OUT}) in series with a perfect voltage source (V_{SOURCE}). This is called the equivalent circuit and it represents the combined effect of all the voltage sources, resistance, capacitance and inductance connected to the output inside the circuit or device. Note that V_{SOURCE} is usually not the same as the supply voltage Vs.

It is normal to use the term 'output impedance' even for simple cases where there is only resistance and the term 'output resistance' could be used instead. In fact it is usually reasonable to assume that an output impedance is just resistance providing the output signal has a low frequency (less than 1kHz).

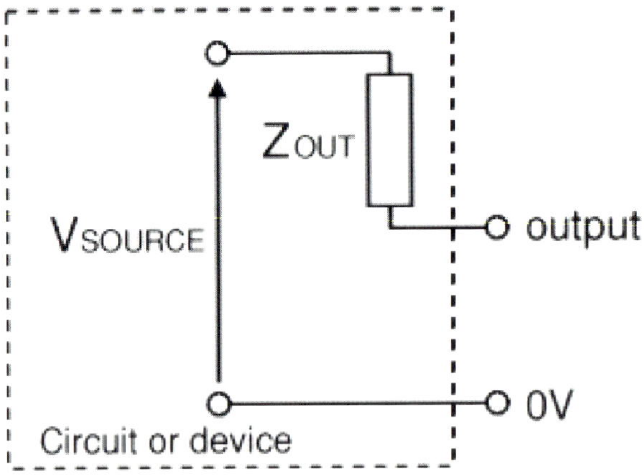

The equivalent circuit of any output

The effects of capacitance and inductance vary with frequency, so if these are present the output impedance will vary with frequency. The effects of capacitance and inductance are generally most significant at high frequencies.

Usually, output impedances should be low, less than a tenth of the load impedance connected to the output. If an output impedance is too high it will be unable to supply a sufficiently strong signal to the load because most of the signal's voltage will be 'lost' inside the circuit driving current through the output impedance Z_{OUT}. The load could be a single component or the input impedance of another circuit.

Low Output Impedance, $Z_{OUT} << Z_{LOAD}$

Most of V_{SOURCE} appears across the load, very little voltage is 'lost' driving the output current through the output impedance. Usually this is the best arrangement.

Matched Impedances, $Z_{OUT} = Z_{LOAD}$

Half of V_{SOURCE} appears across the load, the other half is 'lost' driving the output current through the output impedance. This arrangement is useful in some situations (such as an amplifier driving a loudspeaker) because it delivers maximum power to the load. Note that an equal amount of power is wasted driving the output current through Z_{OUT}, an efficiency of 50%.

High Output Impedance, $Z_{OUT} >> Z_{LOAD}$

Only a small portion of V_{SOURCE} appears across the load, most is 'lost' driving the output current through the output impedance. This arrangement is unsatisfactory.

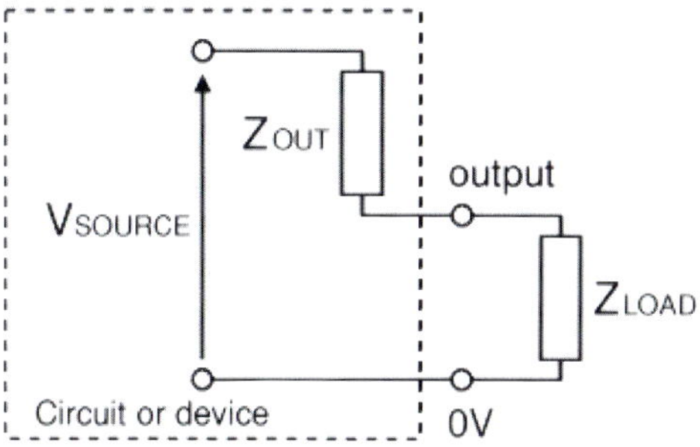

The load can be a single component or the input impedance of another circuit

The Output Impedance of a Voltage Divider

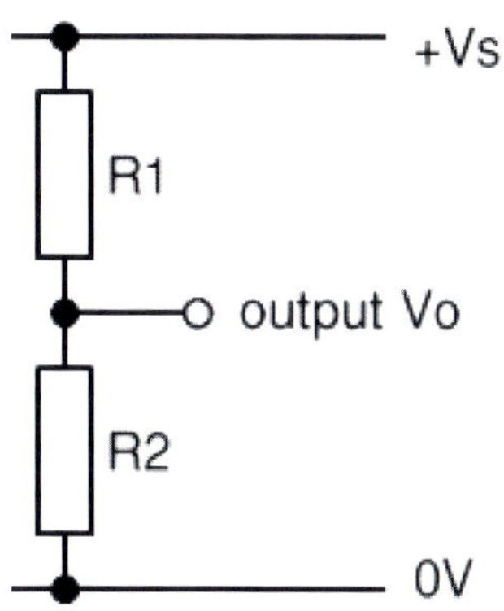

Voltage dividers are widely used in electronics, for example to connect an input transducer such as an LDR to a circuit input.

For successful use the output impedance of the voltage divider should be much smaller than the input impedance of the circuit connected to it.

Ideally the output impedance should be less than a tenth of the input impedance.

In the equivalent circuit of a voltage divider the output impedance is just a resistance and the term 'output resistance' could be used. R_{OUT} is equal to the two resistances (R1 and R2) connected in parallel:

$$\text{Output impedance, } R_{OUT} = \frac{R1 \times R2}{R1 + R2}$$

The voltage source V_{SOURCE} in the equivalent circuit is the value of the output voltage Vo when there is nothing connected to the output (and therefore no output current). It is sometimes called the 'open circuit' voltage.

$$\text{Voltage source, } V_{SOURCE} = \frac{Vs \times R2}{R1 + R2}$$

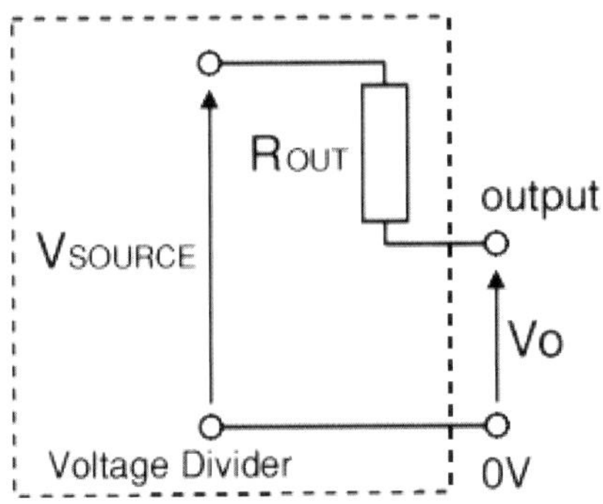

Equivalent circuit of a voltage divider

In most voltage dividers one of the resistors will be aninput transducer such as an LDR. The transducer's resistance varies and this will make both V_{SOURCE} and R_{OUT} vary too. To check that R_{OUT} is sufficiently low you should work out its highest value which will occur when the transducer has its maximum resistance (this applies wherever the transducer is connected in the voltage divider).

For example: If R1 = 10k Ω and R2 is an LDR with maximum resistance 1M Ω, R_{OUT} = 10k × 1M / (10k + 1M) = 9.9k (say 10k Ω). This means it should be connected to a load or input resistance of at least 100k.

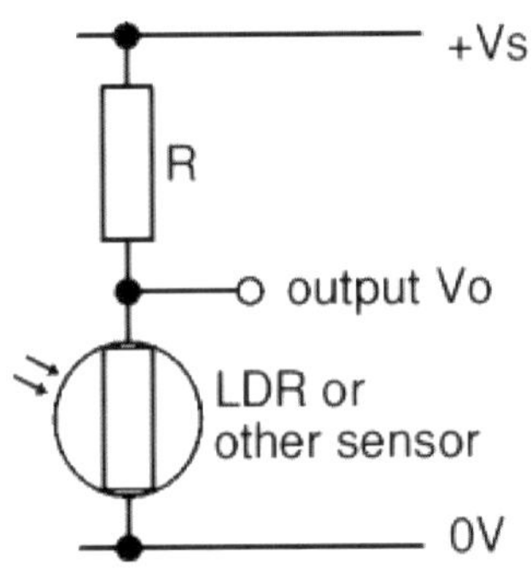

Voltage divider with an LDR

Stress and Strain

The term stress (s) is used to express the loading in terms of force applied to a certain cross-sectional area of an object. From the perspective of loading, stress is the applied force or system of forces that tends to deform a body. From the perspective of what is happening within a material, stress is the internal distribution of forces within a body that balance and react to the loads applied to it. The stress distribution may or may not be uniform, depending on the nature of the loading condition. For example, a bar loaded in pure tension will essentially have a uniform tensile stress distribution. However, a bar loaded in bending will have a stress distribution that changes with distance perpendicular to the normal axis.

Simplifying assumptions are often used to represent stress as a vector quantity for many engineering calculations and for material property determination. The word *"vector"* typically refers to a quantity that has a "magnitude" and a "direction". For example, the stress in an axially loaded bar is simply equal to the applied force divided by the bar's cross-sectional area.

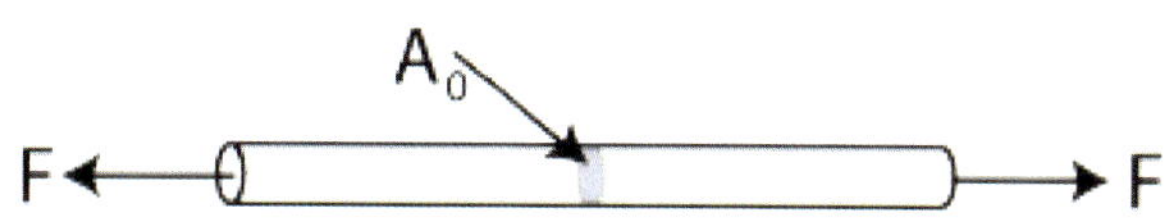

$$\text{Stress, } \sigma = \frac{\text{Force}}{\text{Cross-Sectional Area}} = \frac{F}{A_0}$$

Some common measurements of stress are:

psi = lbs/in^2 (pounds per square inch)

ksi or kpsi = kilopounds/in^2 (one thousand or 10^3 pounds per square inch)

Pa = N/m^2 (Pascals or Newtons per square meter)

kPa = Kilopascals (one thousand or 10^3 Newtons per square meter)

GPa = Gigapascals (one million or 10^6 Newtons per square meter)

*Any metric prefix can be added in front of psi or Pa to indicate the multiplication factor

It must be noted that the stresses in most 2-D or 3-D solids are actually more complex and need be defined more methodically. The internal force acting on a small area of a plane can be resolved into three components: one normal to the plane and two parallel to the plane. The normal force component divided by the area gives the normal stress (s), and parallel force components divided by the area give the shear stress (t).

These stresses are average stresses as the area is finite, but when the area is allowed to approach zero, the stresses become stresses at a point. Since stresses are defined in relation to the plane that passes through the point under consideration, and the number of such planes is infinite, there appear an infinite set of stresses at a point. Fortunately, it can be proven that the stresses on any plane can be computed from the stresses on three orthogonal planes passing through the point. As each plane has three stresses, the stress tensor has nine stress components, which completely describe the state of stress at a point.

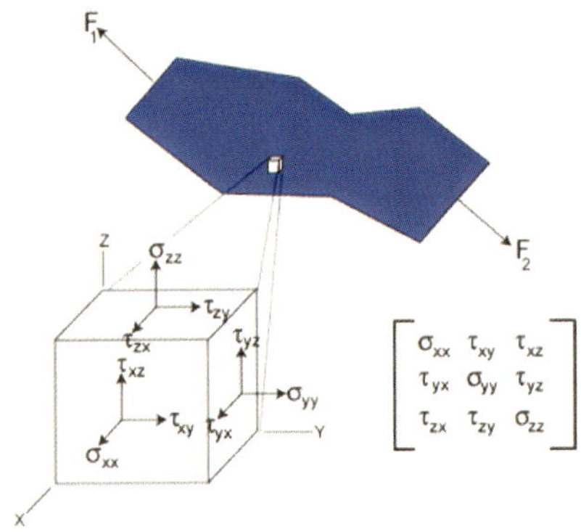

Strain

Strain is the response of a system to an applied stress. When a material is loaded with a force, it produces a stress, which then causes a material to deform. Engineering strain is defined as the amount of deformation in the direction of the applied force divided by the initial length of the material. This results in a unitless number, although it is often left in the unsimplified form, such as inches per inch or meters per meter. For example, the strain in a bar that is being stretched in tension is the amount of elongation or change in length divided by its original length. As in the case of stress, the strain distribution may or may not be uniform in a complex structural element, depending on the nature of the loading condition.

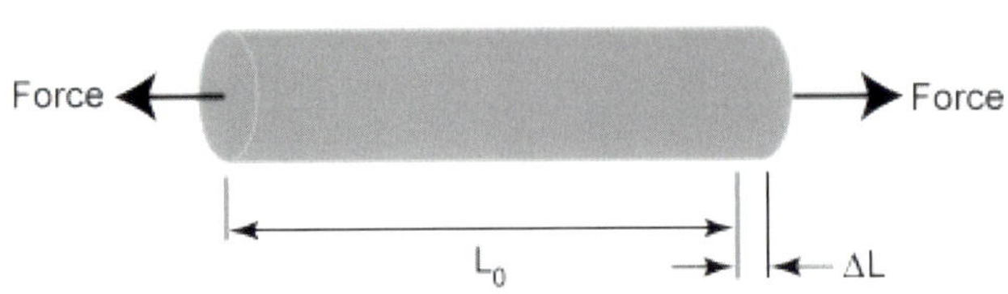

$$\text{Strain} = \frac{\text{Elongation}}{\text{Orininal Length}} = \frac{\Delta L}{L_o}$$

If the stress is small, the material may only strain a small amount and the material will return to its original size after the stress is released. This is called elastic deformation, because like elastic it returns to its unstressed state. Elastic deformation only occurs in a material when stresses are lower than a critical stress called the yield strength. If a material is loaded beyond it elastic limit, the material will remain in a deformed condition after the load is removed. This is called plastic deformation.

Engineering and True Stress and Strain

The discussion focused on *engineering* stress and strain, which use the fixed, undeformed cross-sectional area in the calculations. *True* stress and strain measures account for changes in cross-sectional area by using the *instantaneous* values for the area. The engineering stress-strain curve does not give a true indication of the deformation characteristics of a metal because it is based entirely on the original dimensions of the specimen, and these dimensions change continuously during the testing used to generate the data.

Engineering stress and strain data is commonly used because it is easier to generate the data and the tensile properties are adequate for engineering calculations. When considering the stress-strain curves, however, it should be understood that metals and other materials continues to strain-harden until they fracture and the stress required to produce further deformation also increase.

Stress Concentration

When an axial load is applied to a piece of material with a uniform cross-section, the norm al stress will be uniformly distributed over the cross-section. However, if a hole is drilled in the material, the stress distribution will no longer be uniform. Since the material that has been removed from the hole is no longer available to carry any load, the load must be redistributed over the remaining material. It is not redistributed evenly over the entire remaining cross-sectional area but instead will be redistributed in an uneven pattern that is highest at the edges of the hole as shown in the image. This phenomenon is known as stress concentration.

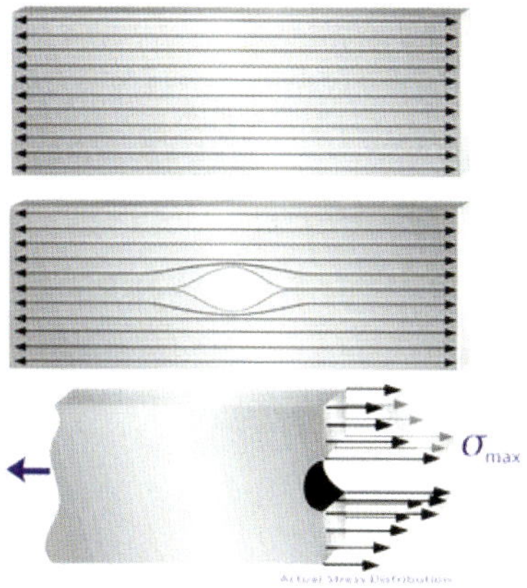

Stress Concentration Due to a Hole

Stress-strain Curve

Suppose that a metal specimen be placed in tension-compression-testing machine. As the axial load is gradually increased in increments, the total elongation over the gauge length is measured at each increment of the load and this is continued until failure of the specimen takes place. Knowing the original cross-sectional area and length of the specimen, the normal stress σ and the strain ε can be obtained. The graph of these quantities with the stress σ along the y-axis and the strain ε along the x-axis is called the stress-strain diagram. The stress-strain diagram differs in form for various materials. The diagram shown below is that for a medium-carbon structural steel.

Metallic engineering materials are classified as either ductile or brittle materials. A ductile material is one having relatively large tensile strains up to the point of rupture like structural steel and aluminum, whereas brittle materials has a relatively small strain up to the point of rupture like cast iron and concrete. An arbitrary strain of 0.05 mm/mm is frequently taken as the dividing line between these two classes.

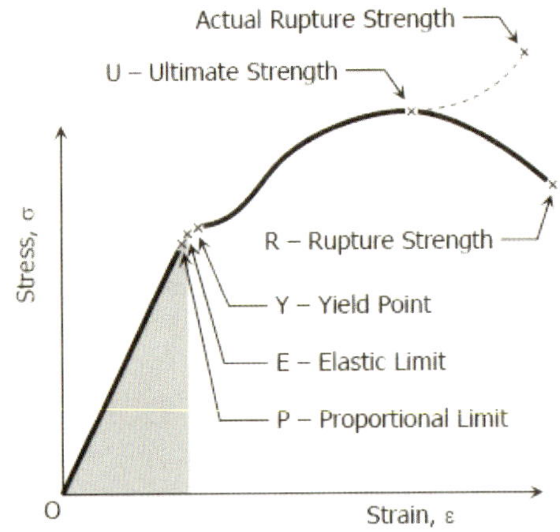

Stress-strain diagram of a medium-carbon structural steel

Proportional Limit (Hooke's Law)

Robert Hooke

From the origin O to the point called proportional limit, the stress-strain curve is a straight line. This linear relation between elongation and the axial force causing was first noticed by Sir Robert Hooke in 1678 and is called Hooke's Law that within the proportional limit, the stress is directly proportional to strain or $\sigma \propto \varepsilon \ or \ \sigma = K\varepsilon$.

The constant of proportionality k is called the Modulus of Elasticity E or Young's Modulus and is equal to the slope of the stress-strain diagram from O to P. Then

$$\sigma = E\varepsilon.$$

Elastic Limit

The elastic limit is the limit beyond which the material will no longer go back to its original shape when the load is removed, or it is the maximum stress that may have developed such that there is no permanent or residual deformation when the load is entirely removed.

Elastic and Plastic Ranges

The region in stress-strain diagram from O to P is called the elastic range. The region from P to R is called the plastic range.

Yield Point

Yield point is the point at which the material will have an appreciable elongation or yielding without any increase in load.

Ultimate Strength

The maximum ordinate in the stress-strain diagram is the ultimate strength or tensile strength.

Rupture Strength

Rupture strength is the strength of the material at rupture. This is also known as the breaking strength.

Modulus of Resilience

Modulus of resilience is the work done on a unit volume of material as the force is gradually increased from O to P, in $N \cdot m/m^3$. This may be calculated as the area under the stress-strain curve from the origin O to up to the elastic limit E (the shaded area in the figure). The resilience of the material is its ability to absorb energy without creating a permanent distortion.

Modulus of Toughness

Modulus of toughness is the work done on a unit volume of material as the force is gradually increased from O to R, in N·m/m³. This may be calculated as the area under the entire stress-strain curve (from O to R). The toughness of a material is its ability to absorb energy without causing it to break.

Working Stress, Allowable Stress, and Factor of Safety

Working stress is defined as the actual stress of a material under a given loading. The maximum safe stress that a material can carry is termed as the allowable stress. The allowable stress should be limited to values not exceeding the proportional limit. However, since proportional limit is difficult to determine accurately, the allowable tress is taken as either the yield point or ultimate strength divided by a factor of safety. The ratio of this strength (ultimate or yield strength) to allowable strength is called the factor of safety.

Beam Theory

The beam theory, also known as Euler-Bernoulli beam theory, is a cornerstone of engineering and a driving force of the Second Industrial Revolution. It is a simplification of the theory of elasticity, which allows the calculation of the deflection and load-carrying characteristics of beams. It also is applicable to small deflections of the beam that has been subjected to lateral loads. The beam theory was first proposed in the year 1750 by Leonhard Euler and Daniel Bernoulli, but was applied on a massive scale only after the construction of the Eiffel Tower or the development of the Ferris wheel, both landmarks in human innovation of the late 19th century. Since the validation of the theory on such large scales, it became established as an important tool in structural and mechanical engineering.

Static Beam Equation

The Euler–Bernoulli equation describes the relationship between the beam's deflection and the applied load:

$$\frac{\mathrm{d}^2}{\mathrm{d}x^2}\left(EI\frac{\mathrm{d}^2w}{\mathrm{d}x^2}\right)=q$$

The curve $w(x)$ describes the deflection of the beam in the z direction at some position x (recall that the beam is modeled as a one-dimensional object). q is a distributed load, in other words a force per unit length (analogous to pressure being a force per area); it may be a function of x, w, or other variables.

Note that E is the elastic modulus and that I is the second moment of area of the beam's cross-section. I must be calculated with respect to the axis which passes through the centroid of the cross-section and which is perpendicular to the applied loading.Explicitly, for a beam whose axis is oriented along x with a loading along z, the beam's cross-section is in the yz plane, and the relevant second moment of area is

$$I = \iint z^2 \, dy \, dz,$$

where it is assumed that the centroid of the cross-section occurs at $y = z = 0$.

Often, the product EI (known as the flexural rigidity) is a constant, so that

$$EI \frac{\mathrm{d}^4 w}{\mathrm{d}x^4} = q(x).$$

This equation, describing the deflection of a uniform, static beam, is used widely in engineering practice. Tabulated expressions for the deflection w for common beam configurations can be found in engineering handbooks. For more complicated situations the deflection can be determined by solving the Euler–Bernoulli equation using techniques such as the "slope deflection method", "moment distribution method", "moment area method, "conjugate beam method", "the principle of virtual work", "direct integration", "Castigliano's method", "Macaulay's method" or the "direct stiffness method".

Sign conventions are defined here since different conventions can be found in the literature. A right-handed coordinate system is used as shown in the figure below. In this figure, the x and z direction of a right-handed coordinate system are shown. Since $\mathrm{e}_z \times \mathrm{e}_x = \mathrm{e}_y$ where $\mathrm{e}_x, \mathrm{e}_y$, and e_z are unit vectors in the direction of the x, y, and z axes respectively, the y axis direction is into the figure. Forces acting in the positive x and z directions are assumed positive. The sign of the bending moment M is positive when the torque vector associated with the bending moment on the right hand side of the section is in the positive y direction (i.e. so that a positive value of M leads to a compressive stress at the bottom fibers). With this choice of bending moment sign convention, in order to have $dM = Qdx$,, it is necessary that Q the shear force acting on the right side of the section be positive in the z direction so as to achieve static equilibrium of moments. To have force equilibrium with $dQ = qdx$, q, the loading intensity must be positive in the minus z direction. In addition to these sign conventions for scalar quantities, we also sometimes use vectors in which the directions of the vectors is made clear through the use of the unit vectors, $\mathrm{e}_x, \mathrm{e}_y$, and e_z.

Successive derivatives of the deflection w have important physical meanings: $\mathrm{d}w/\mathrm{d}x$ is the slope of the beam,

$$M = -EI \frac{d^2 w}{dx^2}$$

is the bending moment in the beam, and

$$Q = -\frac{d}{dx}\left(EI\frac{d^2w}{dx^2} \right)$$

is the shear force in the beam.

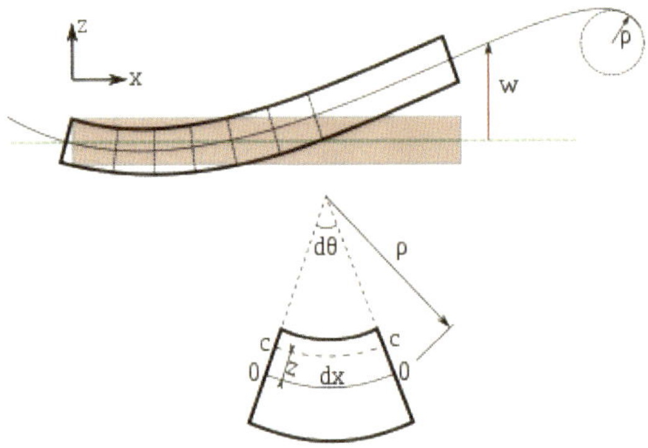

Bending of an Euler–Bernoulli beam. Each cross-section of the beam is at 90 degrees to the neutral axis.

The stresses in a beam can be calculated from the above expressions after the deflection due to a given load has been determined.

Derivation of Bending Moment Equation

The length of the neutral axis in the figure is $\rho d\theta$. The length of a fiber with a radial distance, e, below the neutral axis is $(\rho + e)d\theta$. Therefore, the strain of this fiber is

$$\frac{(\rho + e - \rho)\, d\theta}{\rho\, d\theta} = \frac{e}{\rho}.$$

The stress of this fiber is $E\frac{e}{\rho}$ where E is the elastic modulus in accordance with Hooke's Law. The differential force vector, $d\mathrm{F}$, resulting from this stress is given by,

$$d\mathrm{F} = E\frac{e}{\rho}dA\mathrm{e}_{\mathbf{x}}.$$

This is the differential force vector exerted on the right hand side of the section shown in the figure. We know that it is in the e_x direction since the figure clearly shows that the fibers in the lower half are in tension. dA is the differential element of area at the location of the fiber. The differential bending moment vector, $d\mathrm{M}$ associated with $d\mathrm{F}$ is given by

$$d\mathrm{M} = -e\mathrm{e}_z \times d\mathrm{F} = -\mathrm{e}_{\mathbf{y}}E\frac{e^2}{\rho}dA.$$

This expression is valid for the fibers in the lower half of the beam. The expression for the fibers in the upper half of the beam will be similar except that the moment arm vector will be in the positive z direction and the force vector will be in the -x direction since the upper fibers are in compression. But the resulting bending moment vector will still be in the -y direction since $e_z \times -e_x = -e_y$. Therefore, we integrate over the entire cross section of the beam and get for $\mathbf{M}$ the bending moment vector exerted on the right cross section of the beam the expression

$$\mathbf{M} = \int d\mathbf{M} = -\mathbf{e}_y \frac{E}{\rho} \int e^2 \, dA = -\mathbf{e}_y \frac{EI}{\rho},$$

where I is the second moment of area. From calculus, we know that when $\frac{dw}{dx}$ is small as it is for an Euler–Bernoulli beam, $\frac{1}{\rho} = \frac{d^2w}{dx^2}$ (ρ is the radius of curvature). Therefore,

$$\mathbf{M} = -\mathbf{e}_y EI \frac{d^2 w}{dx^2} = -\mathbf{e}_y EI \frac{d^2 z}{dx^2}.$$

Dynamic Beam Equation

The dynamic beam equation is the Euler–Lagrange equation for the following action

$$S = \int_0^L \left[\frac{1}{2}\mu\left(\frac{\partial w}{\partial t}\right)^2 - \frac{1}{2}EI\left(\frac{\partial^2 w}{\partial x^2}\right)^2 + q(x)w(x,t) \right] dx.$$

The first term represents the kinetic energy where μ is the mass per unit length; the second one represents the potential energy due to internal forces (when considered with a negative sign) and the third term represents the potential energy due to the external load $q(x)$. The Euler–Lagrange equation is used to determine the function that minimizes the functional S. For a dynamic Euler–Bernoulli beam, the Euler–Lagrange equation is

$$\frac{\partial^2}{\partial x^2}\left(EI\frac{\partial^2 w}{\partial x^2}\right) = -\mu\frac{\partial^2 w}{\partial t^2} + q(x)$$

Derivation of Euler–Lagrange Equation for Beams

Since the Lagrangian is

$$\mathcal{L} := \tfrac{1}{2}\mu\left(\frac{\partial w}{\partial t}\right)^2 - \tfrac{1}{2}EI\left(\frac{\partial^2 w}{\partial x^2}\right)^2 + q(x)w(x,t) = \tfrac{\mu}{2}\dot{w}^2 - \tfrac{EI}{2}w_{xx}^2 + qw \equiv \mathcal{L}(x,t,w,\dot{w},w_{xx})$$

the corresponding Euler–Lagrange equation is

$$\frac{\partial \mathcal{L}}{\partial w} - \frac{\partial}{\partial t}\left(\frac{\partial \mathcal{L}}{\partial \dot{w}}\right) + \frac{\partial^2}{\partial x^2}\left(\frac{\partial \mathcal{L}}{\partial w_{xx}}\right) = 0$$

Now,

$$\frac{\partial \mathcal{L}}{\partial w} = q \; ; \; \frac{\partial \mathcal{L}}{\partial \dot{w}} = \mu \dot{w} \; ; \; \frac{\partial \mathcal{L}}{\partial w_{xx}} = -EIw_{xx} \; .$$

Plugging into the Euler–Lagrange equation gives

$$q - \mu \ddot{w} - (EIw_{xx})_{xx} = 0$$

or,

$$\frac{\partial^2}{\partial x^2}\left(EI \frac{\partial^2 w}{\partial x^2} \right) = -\mu \frac{\partial^2 w}{\partial t^2} + q$$

which is the governing equation for the dynamics of an Euler–Bernoulli beam.

When the beam is homogeneous, E and I are independent of x, and the beam equation is simpler:

$$EI \frac{\partial^4 w}{\partial x^4} = -\mu \frac{\partial^2 w}{\partial t^2} + q.$$

Free Vibration

In the absence of a transverse load, q, we have the free vibration equation. This equation can be solved using a Fourier decomposition of the displacement into the sum of harmonic vibrations of the form

$$w(x,t) = \mathrm{Re}[\hat{w}(x) e^{-i\omega t}]$$

where ω is the frequency of vibration. Then, for each value of frequency, we can solve an ordinary differential equation

$$EI \frac{\mathrm{d}^4 \hat{w}}{\mathrm{d}x^4} + \mu \omega^2 \hat{w} = 0.$$

The general solution of the above equation is

$$\hat{w} = A_1 \cosh(\beta x) + A_2 \sinh(\beta x) + A_3 \cos(\beta x) + A_4 \sin(\beta x) \quad \text{with} \quad \beta := \left(\frac{\mu \omega^2}{EI} \right)^{1/4}$$

where A_1, A_2, A_3, A_4 are constants. These constants are unique for a given set of boundary conditions. However, the solution for the displacement is not unique and depends on the frequency. These solutions are typically written as

$$\hat{w}_n = A_1 \cosh(\beta_n x) + A_2 \sinh(\beta_n x) + A_3 \cos(\beta_n x) + A_4 \sin(\beta_n x) \quad \text{with} \quad \beta_n := \left(\frac{\mu\omega_n^2}{EI}\right)^{1/4}.$$

The quantities ω_n are called the natural frequencies of the beam. Each of the displacement solutions is called a mode and the shape of the displacement curve is called the mode shape.

Example: Cantilevered Beam

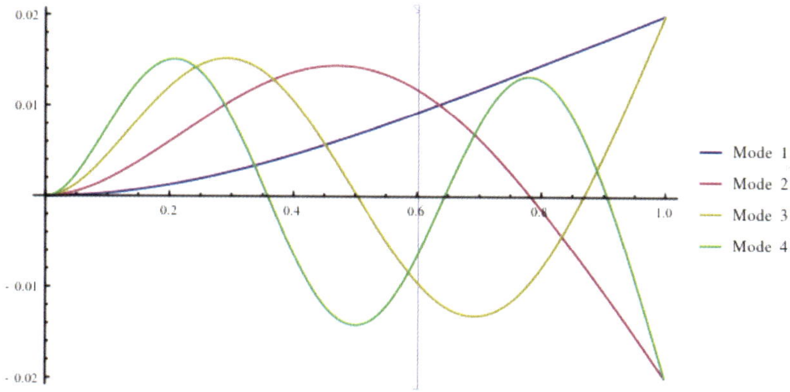

Mode shapes for the first four modes of a vibrating cantilever beam.

The boundary conditions for a cantilevered beam of length L (fixed at $x = 0$) are

$$\hat{w}_n = 0\,, \quad \frac{d\hat{w}_n}{dx} = 0 \quad \text{at } x = 0$$

$$\frac{d^2\hat{w}_n}{dx^2} = 0\,, \quad \frac{d^3\hat{w}_n}{dx^3} = 0 \quad \text{at } x = L.$$

If we apply these conditions, non-trivial solutions are found to exist only if $\cosh(\beta_n L)\cos(\beta_n L) + 1 = 0$. This nonlinear equation can be solved numerically. The first few roots are $\beta_1 L/\pi =$ 0.59686...,$\beta_2 L/\pi =$ 1.49418...,$\beta_3 L/\pi =$ 2.50025...,$\beta_4 L/\pi =$ 3.49999..., ...

The corresponding natural frequencies of vibration are

$$\omega_1 = \beta_1^2 \sqrt{\frac{EI}{\mu}} = \frac{3.5161}{L^2}\sqrt{\frac{EI}{\mu}}\,, \ldots$$

The boundary conditions can also be used to determine the mode shapes from the solution for the displacement:

$$\hat{w}_n = A_1\left[(\cosh\beta_n x - \cos\beta_n x) + \frac{\cos\beta_n L + \cosh\beta_n L}{\sin\beta_n L + \sinh\beta_n L}(\sin\beta_n x - \sinh\beta_n x)\right]$$

The unknown constant (actually constants as there is one for each n), A_1, which in general is complex, is determined by the initial conditions at $t=0$ on the velocity and displacements of the beam. Typically a value of $A_1=1$ is used when plotting mode shapes. Solutions to the undampened forced problem have unbounded displacements when the driving frequency matches a natural frequency ω_n, i.e., the beam can resonate. The natural frequencies of a beam therefore correspond to the frequencies at which resonance can occur.

Example: Unsupported (Free-free) Beam

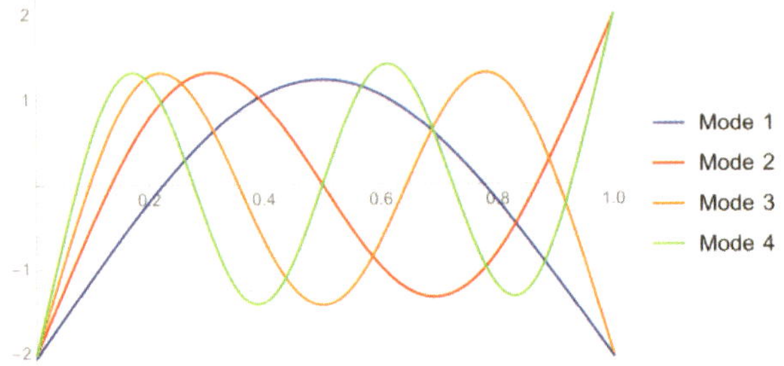

The first four modes of a vibrating free-free Euler-Bernoulli beam.

A free-free beam is a beam without any supports. The boundary conditions for a free beam of length L extending from x=0 to x=L is given by:

$$\frac{d^2\hat{w}_n}{dx^2}=0\,,\ \frac{d^3\hat{w}_n}{dx^3}=0\quad \text{at } x=0 \text{ and } L.$$

If we apply these conditions, non-trivial solutions are found to exist only if $\cosh(\beta_n L)\cos(\beta_n L)+1=0$.

This nonlinear equation can be solved numerically. The first few roots are $\beta_1 L/\pi=$ 1.50562...,$\beta_2 L/\pi=$ 2.49975...,$\beta_3 L/\pi=$ 3.50001...,$\beta_4 L/\pi=$ 4.50000... The corresponding natural frequencies of vibration are:

$$\omega_1=\beta_1^2\sqrt{\frac{EI}{\mu}}=\frac{22.3733}{L^2}\sqrt{\frac{EI}{\mu}},\ldots$$

The boundary conditions can also be used to determine the mode shapes from the solution for the displacement:

$$\hat{w}_n=A_1\left[(\sin\beta_n x+\sinh\beta_n x)-\frac{\sin\beta_n L-\sinh\beta_n L}{\cos\beta_n L-\cosh\beta_n L}(\cos\beta_n x+\cosh\beta_n x)\right]$$

As with the cantilevered beam, the unknown constants are determined by the initial conditions at $t=0$ on the velocity and displacements of the beam. Also, solutions to the undampened forced problem have unbounded displacements when the driving frequency matches a natural frequency ω_n.

Stress

Besides deflection, the beam equation describes forces and moments and can thus be used to describe stresses. For this reason, the Euler–Bernoulli beam equation is widely used in engineering, especially civil and mechanical, to determine the strength (as well as deflection) of beams under bending.

Both the bending moment and the shear force cause stresses in the beam. The stress due to shear force is maximum along the neutral axis of the beam (when the width of the beam, t, is constant along the cross section of the beam; otherwise an integral involving the first moment and the beam's width needs to be evaluated for the particular cross section), and the maximum tensile stress is at either the top or bottom surfaces. Thus the maximum principal stress in the beam may be neither at the surface nor at the center but in some general area. However, shear force stresses are negligible in comparison to bending moment stresses in all but the stockiest of beams as well as the fact that stress concentrations commonly occur at surfaces, meaning that the maximum stress in a beam is likely to be at the surface.

Simple or Symmetrical Bending

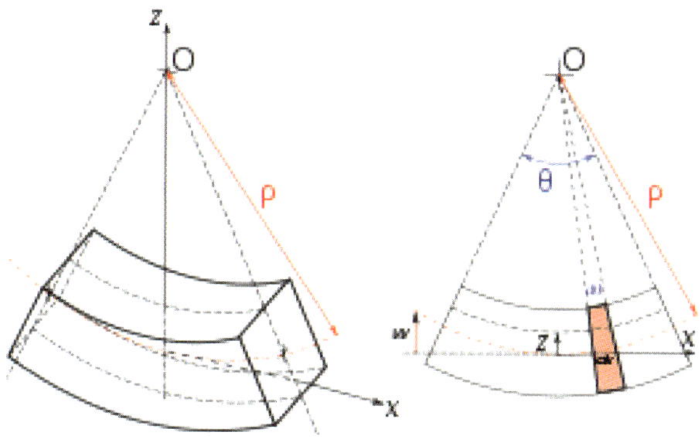

Element of a bent beam: the fibers form concentric arcs, the top fibers are compressed and bottom fibers stretched.

For beam cross-sections that are symmetrical about a plane perpendicular to the neutral plane, it can be shown that the tensile stress experienced by the beam may be expressed as:

$$\sigma = \frac{Mz}{I} = -zE\frac{d^2w}{dx^2}.$$

Here, z is the distance from the neutral axis to a point of interest; and M is the bending moment. Note that this equation implies that pure bending (of positive sign) will cause zero stress at the neutral axis, positive (tensile) stress at the "top" of the beam, and negative (compressive) stress at the bottom of the beam; and also implies that the maximum stress will be at the top surface and the minimum at the bottom. This bending stress may be superimposed with axially applied stresses, which will cause a shift in the neutral (zero stress) axis.

Maximum Stresses at a Cross-section

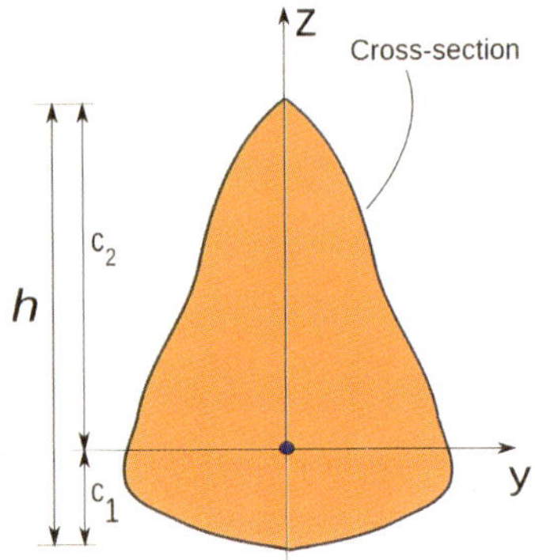

Quantities used in the definition of the section modulus of a beam.

The maximum tensile stress at a cross-section is at the location $z = c_1$ and the maximum compressive stress is at the location $z = -c_2$ where the height of the cross-section is $h = c_1 + c_2$.. These stresses are

$$\sigma_1 = \frac{Mc_1}{I} = \frac{M}{S_1} \; ; \; \sigma_2 = -\frac{Mc_2}{I} = -\frac{M}{S_2}$$

The quantities S_1, S_2 are the section moduli and are defined as

$$S_1 = \frac{I}{c_1} \; ; \; S_2 = \frac{I}{c_2}$$

The section modulus combines all the important geometric information about a beam's section into one quantity. For the case where a beam is doubly symmetric, and we have $c_1 = c_2$ one section modulus $S = I / c$.

Strain in an Euler–Bernoulli Beam

We need an expression for the strain in terms of the deflection of the neutral surface to relate the stresses in an Euler–Bernoulli beam to the deflection. To obtain that expression we use the assumption that normals to the neutral surface remain normal during the deformation and that deflections are small. These assumptions imply that the beam bends into an arc of a circle of radius ρ and that the neutral surface does not change in length during the deformation.

Let dx be the length of an element of the neutral surface in the undeformed state. For small deflections, the element does not change its length after bending but deforms into an arc of a circle of radius ρ. If $d\theta$ is the angle subtended by this arc, then $dx = \rho d\theta$.

Let us now consider another segment of the element at a distance z above the neutral surface. The initial length of this element is $\mathrm{d}x$. However, after bending, the length of the element becomes $\mathrm{d}x' = (\rho - z)\mathrm{d}\theta = \mathrm{d}x - z\,\mathrm{d}\theta$. The strain in that segment of the beam is given by

$$\varepsilon_x = \frac{\mathrm{d}x' - \mathrm{d}x}{\mathrm{d}x} = -\frac{z}{\rho} = -\kappa z$$

where κ is the curvature of the beam. This gives us the axial strain in the beam as a function of distance from the neutral surface. However, we still need to find a relation between the radius of curvature and the beam deflection w.

Relation between Curvature and Beam Deflection

Let P be a point on the neutral surface of the beam at a distance x from the origin of the (x,z) coordinate system. The slope of the beam is approximately equal to the angle made by the neutral surface with the x -axis for the small angles encountered in beam theory. Therefore, with this approximation,

$$\theta(x) = \frac{\mathrm{d}w}{\mathrm{d}x}$$

Therefore, for an infinitesimal element $\mathrm{d}x$, the relation $\mathrm{d}x = \rho\,\mathrm{d}\theta$ can be written as

$$\frac{1}{\rho} = \frac{\mathrm{d}\theta}{\mathrm{d}x} = \frac{\mathrm{d}^2 w}{\mathrm{d}x^2} = \kappa$$

Hence the strain in the beam may be expressed as

$$\varepsilon_x = -z\kappa$$

Stress-strain Relations

For a homogeneous isotropic linear elastic material, the stress is related to the strain by $\sigma = E\varepsilon$, where E is the Young's modulus. Hence the stress in an Euler–Bernoulli beam is given by

$$\sigma_x = -zE\frac{\mathrm{d}^2 w}{\mathrm{d}x^2}$$

Note that the above relation, when compared with the relation between the axial stress and the bending moment, leads to

$$M = -EI\frac{\mathrm{d}^2 w}{\mathrm{d}x^2}$$

Since the shear force is given by $Q = \mathrm{d}M / \mathrm{d}x$,, we also have

$$Q = -EI\frac{\mathrm{d}^3 w}{\mathrm{d}x^3}$$

Boundary Considerations

The beam equation contains a fourth-order derivative in x. To find a unique solution $w(x,t)$ we need four boundary conditions. The boundary conditions usually model *supports*, but they can also model point loads, distributed loads and moments. The *support* or displacement boundary conditions are used to fix values of displacement (w) and rotations $(\mathrm{d}w/\mathrm{d}x)$ on the boundary. Such boundary conditions are also called Dirichlet boundary conditions. Load and moment boundary conditions involve higher derivatives of w and represent momentum flux. Flux boundary conditions are also called Neumann boundary conditions.

As an example consider a cantilever beam that is built-in at one end and free at the other as shown in the adjacent figure. At the built-in end of the beam there cannot be any displacement or rotation of the beam. This means that at the left end both deflection and slope are zero. Since no external bending moment is applied at the free end of the beam, the bending moment at that location is zero. In addition, if there is no external force applied to the beam, the shear force at the free end is also zero.

Taking the x coordinate of the left end as 0 and the right end as L (the length of the beam), these statements translate to the following set of boundary conditions (assume EI is a constant):

$$w|_{x=0} = 0 \quad ; \quad \frac{\partial w}{\partial x}\Big|_{x=0} = 0 \qquad \text{(fixed end)}$$

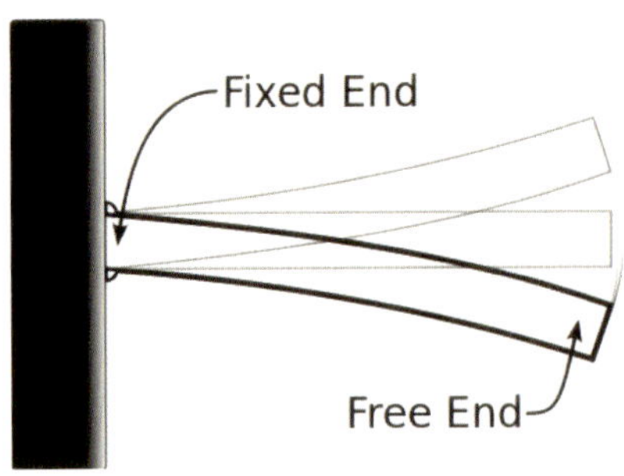

A cantilever beam.

$$\frac{\partial^2 w}{\partial x^2}\Big|_{x=L} = 0 \quad ; \quad \frac{\partial^3 w}{\partial x^3}\Big|_{x=L} = 0 \qquad \text{(free end)}$$

A simple support (pin or roller) is equivalent to a point force on the beam which is adjusted in such a way as to fix the position of the beam at that point. A fixed support or clamp, is equivalent to the combination of a point force and a point torque which is adjusted in such a way as to fix both the position and slope of the beam at that point.

Point forces and torques, whether from supports or directly applied, will divide a beam into a set of segments, between which the beam equation will yield a continuous solution, given four boundary conditions, two at each end of the segment. Assuming that the product EI is a constant, and defining $\lambda = F / EI$ where F is the magnitude of a point force, and $\tau = M / EI$ where M is the magnitude of a point torque, the boundary conditions appropriate for some common cases is given in the table below. The change in a particular derivative of w across the boundary as x increases is denoted by Δ followed by that derivative. For example, $\Delta w'' = w''(x+) - w''(x-)$ where $w''(x+)$ is the value of w'' at the lower boundary of the upper segment, while $w''(x-)$ is the value of w'' at the upper boundary of the lower segment. When the values of the particular derivative are not only continuous across the boundary, but fixed as well, the boundary condition is written e.g., $\Delta w'' = 0^*$ which actually constitutes two separate equations (e.g., $w''(x-) = w''(x+) =$ fixed).

Boundary	w'''	w''	w'	w
Clamp			$\Delta w' = 0^*$	$\Delta w = 0^*$
Simple support		$\Delta w'' = 0$	$\Delta w' = 0$	$\Delta \quad =$
Point force	$\Delta w''' = \lambda$	$\Delta w'' = 0$	$\Delta w' = 0$	$\Delta w = 0$
Point torque	$\Delta w''' = 0$	$\Delta w'' = \tau$	$\Delta w' = 0$	$\Delta w = 0$
Free end	$w''' = 0$	$w'' = 0$		
Clamp at end			w' *fixed*	w *fixed*
Simply supported end		$w'' = 0$		w *fixed*
Point force at end	$w''' = \pm\lambda$	$w'' = 0$		
Point torque at end	$w''' = 0$	$w'' = \pm\tau$		

Note that in the first cases, in which the point forces and torques are located between two segments, there are four boundary conditions, two for the lower segment, and two for the upper. When forces and torques are applied to one end of the beam, there are two boundary conditions given which apply at that end. The sign of the point forces and torques at an end will be positive for the lower end, negative for the upper end.

Loading Considerations

Applied loads may be represented either through boundary conditions or through the function $q(x,t)$ which represents an external distributed load. Using distributed loading is often favorable for simplicity. Boundary conditions are, however, often used to model loads depending on context; this practice being especially common in vibration analysis.

By nature, the distributed load is very often represented in a piecewise manner, since in practice a load isn't typically a continuous function. Point loads can be modeled with help of the Dirac delta function. For example, consider a static uniform cantilever beam of length L with an upward point load F applied at the free end. Using boundary conditions, this may be modeled in two ways. In the first approach, the applied point load is approximated by a shear force applied at the free end. In that case the governing equation and boundary conditions are:

$$EI\frac{d^4w}{dx^4}=0$$

$$w|_{x=0}=0 \;;\quad \frac{dw}{dx}\Big|_{x=0}=0 \;;\quad \frac{d^2w}{dx^2}\Big|_{x=L}=0 \;;\quad -EI\frac{d^3w}{dx^3}\Big|_{x=L}=F$$

Alternatively we can represent the point load as a distribution using the Dirac function. In that case the equation and boundary conditions are

$$EI\frac{d^4w}{dx^4}=F\delta(x-L)$$

$$w|_{x=0}=0 \;;\quad \frac{dw}{dx}\Big|_{x=0}=0 \;;\quad \frac{d^2w}{dx^2}\Big|_{x=L}=0$$

Note that shear force boundary condition (third derivative) is removed, otherwise there would be a contradiction. These are equivalent boundary value problems, and both yield the solution

$$w=\frac{F}{6EI}(3Lx^2-x^3)\,.$$

The application of several point loads at different locations will lead to $w(x)$ being a piecewise function. Use of the Dirac function greatly simplifies such situations; otherwise the beam would have to be divided into sections, each with four boundary conditions solved separately. A well-organized family of functions called Singularity functions are often used as a shorthand for the Dirac function, its derivative, and its antiderivatives.

Dynamic phenomena can also be modeled using the static beam equation by choosing appropriate forms of the load distribution. As an example, the free vibration of a beam can be accounted for by using the load function:

$$q(x,t)=\mu\frac{\partial^2 w}{\partial t^2}$$

where μ is the linear mass density of the beam, not necessarily a constant. With this time-dependent loading, the beam equation will be a partial differential equation:

$$\frac{\partial^2}{\partial x^2}\left(EI\frac{\partial^2 w}{\partial x^2}\right) = -\mu\frac{\partial^2 w}{\partial t^2}.$$

Another interesting example describes the deflection of a beam rotating with a constant angular frequency of ω:

$$q(x) = \mu\omega^2 w(x)$$

This is a centripetal force distribution. Note that in this case, q is a function of the displacement (the dependent variable), and the beam equation will be an autonomous ordinary differential equation.

Examples

Three-point Bending

The three point bending test is a classical experiment in mechanics. It represents the case of a beam resting on two roller supports and subjected to a concentrated load applied in the middle of the beam. The shear is constant in absolute value: it is half the central load, P / 2. It changes sign in the middle of the beam. The bending moment varies linearly from one end, where it is 0, and the center where its absolute value is PL / 4, is where the risk of rupture is the most important. The deformation of the beam is described by a polynomial of third degree over a half beam (the other half being symmetrical). The bending moments (M), shear forces (Q), and deflections (w) for a beam subjected to a central point load and an asymmetric point load are given in the table below.

Distribution	Max. value	
Simply supported beam with central load		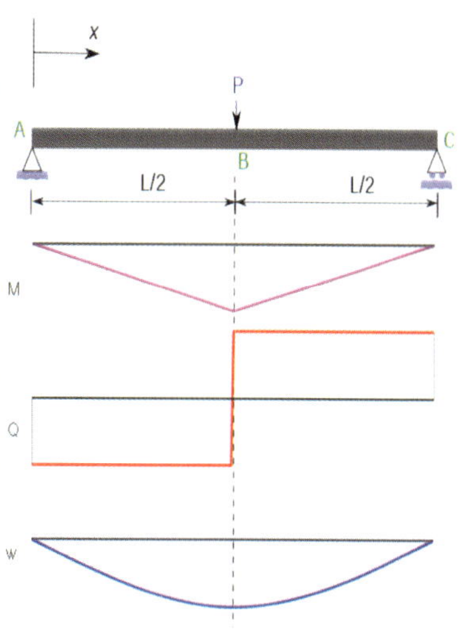
$M(x)=\begin{cases}\frac{Px}{2}, & \text{for } 0\le x\le \frac{L}{2}\\ \frac{P(L-x)}{2}, & \text{for } \frac{L}{2}< x\le L\end{cases}$	$M_{L/2}=\frac{PL}{4}$	
$Q(x)=\begin{cases}\frac{P}{2}, & \text{for } 0\le x\le \frac{L}{2}\\ \frac{-P}{2}, & \text{for } \frac{L}{2}< x\le L\end{cases}$	$\lvert Q_0\rvert=\lvert Q_L\rvert=\frac{P}{2}$	
$w(x)=\begin{cases}-\frac{Px(4x^2-3L^2)}{48EI}, & \text{for } 0\le x\le \frac{L}{2}\\ \frac{P(x-L)(L^2-8Lx+4x^2)}{48EI}, & \text{for } \frac{L}{2}< x\le L\end{cases}$	$w_{L/2}=\frac{PL^3}{48EI}$	

Simply supported beam with asymmetric load		
$M(x)=\begin{cases}\frac{Pbx}{L}, & \text{for } 0\le x\le a\\ \frac{Pbx}{L}-P(x-a)=\frac{Pa(L-x)}{L}, & \text{for } a<x\le L\end{cases}$	$M_B=\frac{Pab}{L}$	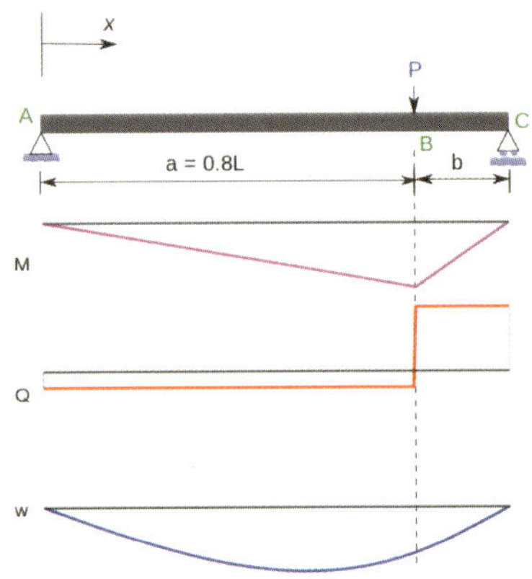
$Q(x)=\begin{cases}\frac{Pb}{L}, & \text{for } 0\le x\le a\\ \frac{Pb}{L}-P, & \text{for } a<x\le L\end{cases}$	$Q_A=\frac{Pb}{L}$ $Q_C=\frac{Pa}{L}$	
$w(x)=\begin{cases}\frac{Pbx(L^2-b^2-x^2)}{6LEI}, & 0\le x\le a\\ \frac{Pbx(L^2-b^2-x^2)}{6LEI}+\frac{P(x-a)^3}{6EI}, & a<x\le L\end{cases}$	$w_{max}=\frac{\sqrt{3}Pb(L^2-b^2)^{\frac{3}{2}}}{27LEI}$ at $x=\sqrt{\frac{L^2-b^2}{3}}$	

Cantilever Beams

Another important class of problems involves cantilever beams. The bending moments (M), shear forces (Q), and deflections (w) for a cantilever beam subjected to a point load at the free end and a uniformly distributed load are given in the table below.

Distribution	Max. value	
Cantilever beam with end load		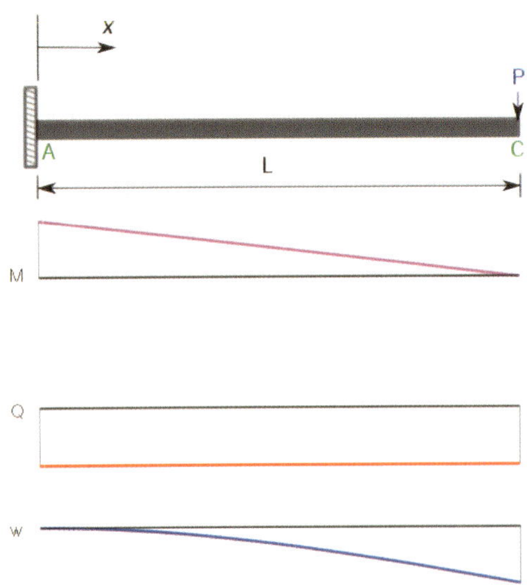
$M(x)=P(L-x)$	$M_A=PL$	
$Q(x)=P$	$Q_{max}=P$	
$w(x)=\frac{Px^2(3L-x)}{6EI}$	$w_C=\frac{PL^3}{3EI}$	

Cantilever beam with uniformly distributed load	
$M(x) = -\frac{q(L^2 - 2Lx + x^2)}{2}$	$M_A = \frac{qL^2}{2}$
$Q(x) = q(L - x),$	$Q_A = qL$
$w(x) = \frac{qx^2(6L^2 - 4Lx + x^2)}{24EI}$	$w_C = \frac{qL^4}{8EI}$

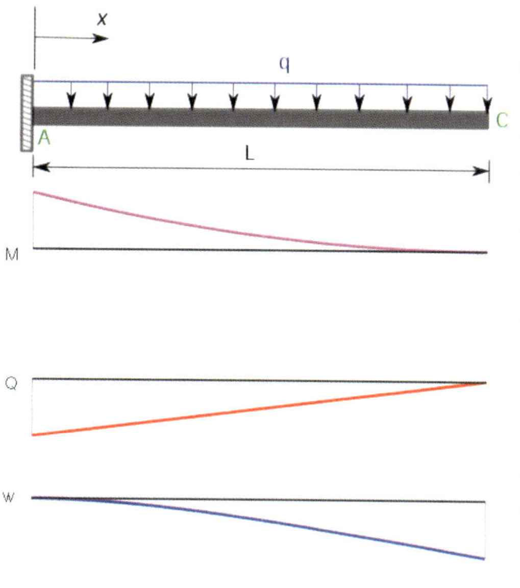

Statically Indeterminate Beams

The bending moments and shear forces in Euler–Bernoulli beams can often be determined directly using static balance of forces and moments. However, for certain boundary conditions, the number of reactions can exceed the number of independent equilibrium equations. Such beams are called *statically indeterminate.*

The built-in beams shown in the figure below are statically indeterminate. To determine the stresses and deflections of such beams, the most direct method is to solve the Euler–Bernoulli beam equation with appropriate boundary conditions. But direct analytical solutions of the beam equation are possible only for the simplest cases. Therefore, additional techniques such as linear superposition are often used to solve statically indeterminate beam problems.

The superposition method involves adding the solutions of a number of statically determinate problems which are chosen such that the boundary conditions for the sum of the individual problems add up to those of the original problem.

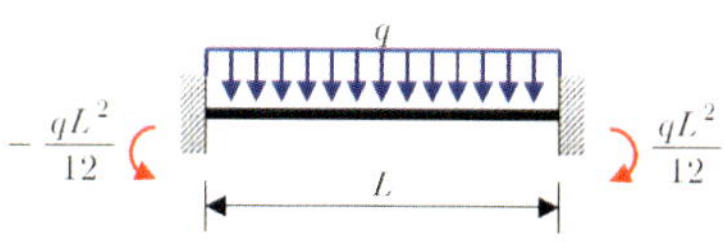

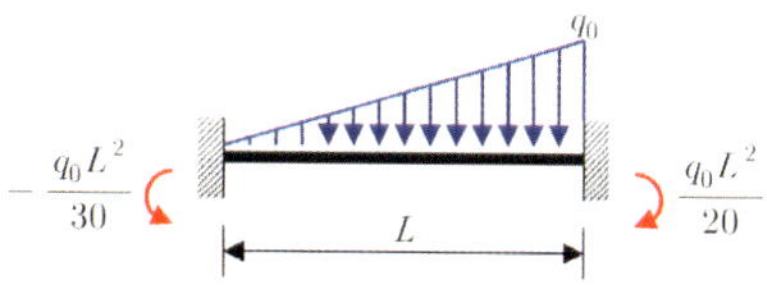

(a) Uniformly distributed load q.	(b) Linearly distributed load with maximum q_o
$M_{max} = \frac{qL^2}{12}$; $w_{max} = \frac{qL^4}{384EI}$	$M_{max} = \frac{qL^2}{300}[3\sqrt{30} - 10]$; $w_{max} = \frac{qL^4}{2500EI}[75 - 7\sqrt{105}]$

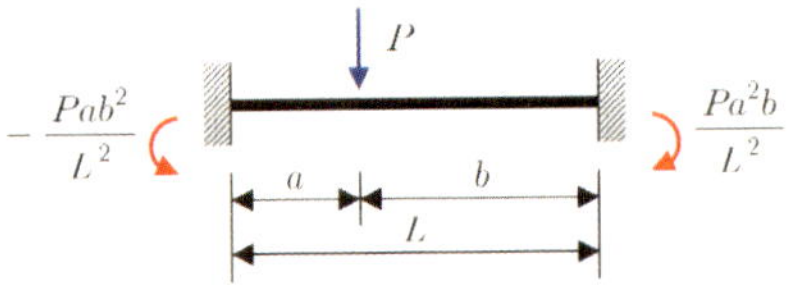	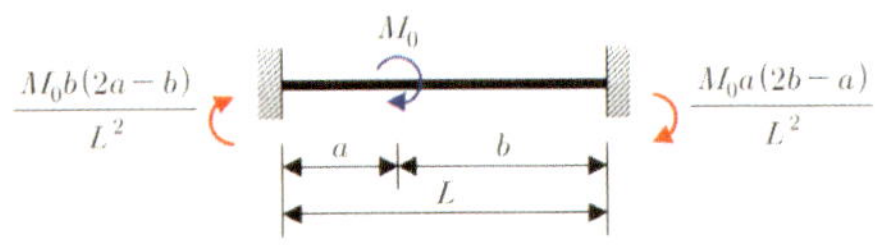
(c) Concentrated load P	(d) Moment M_o

Another commonly encountered statically indeterminate beam problem is the cantilevered beam with the free end supported on a roller. The bending moments, shear forces, and deflections of such a beam are listed below.

Distribution	Max. value
$M(x) = -\frac{q}{8}(L^2 - 5Lx + 4x^2)$	$M_B = -\frac{9qL^2}{128}$ at x $= \frac{5L}{8}$ M_A $\frac{qL^2}{8}$
$Q(x) = -\frac{q}{8}(8x - 5L)$	$Q_A = -\frac{5qL}{8}$
$w(x) = \frac{qx^2}{48EI}(3L^2 - 5Lx + 2x^2)$	$w_{\max} = \frac{(39-55\sqrt{33})}{65{,}536}\frac{qL^4}{EI}$ at $x = \frac{15-\sqrt{33}}{16}L$

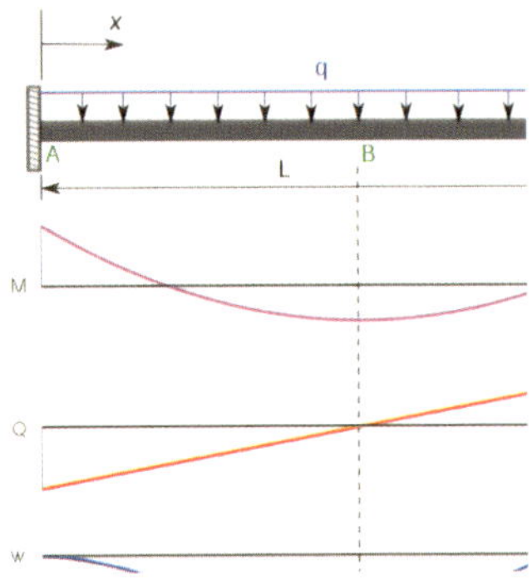

Extensions

The kinematic assumptions upon which the Euler–Bernoulli beam theory is founded allow it to be extended to more advanced analysis. Simple superposition allows for three-dimensional transverse loading. Using alternative constitutive equations can allow for viscoelastic or plastic beam deformation. Euler–Bernoulli beam theory can also be extended to the analysis of curved beams, beam buckling, composite beams, and geometrically nonlinear beam deflection.

Euler–Bernoulli beam theory does not account for the effects of transverse shear strain. As a result, it underpredicts deflections and overpredicts natural frequencies. For thin beams (beam length to thickness ratios of the order 20 or more) these effects are of minor importance. For thick beams, however, these effects can be significant. More advanced beam theories such as the Timoshenko beam theory (developed by the Russian-born scientist Stephen Timoshenko) have been developed to account for these effects.

Large Deflections

The original Euler–Bernoulli theory is valid only for infinitesimal strains and small rotations. The theory can be extended in a straightforward manner to problems involving moderately large rotations provided that the strain remains small by using the von Kármán strains.

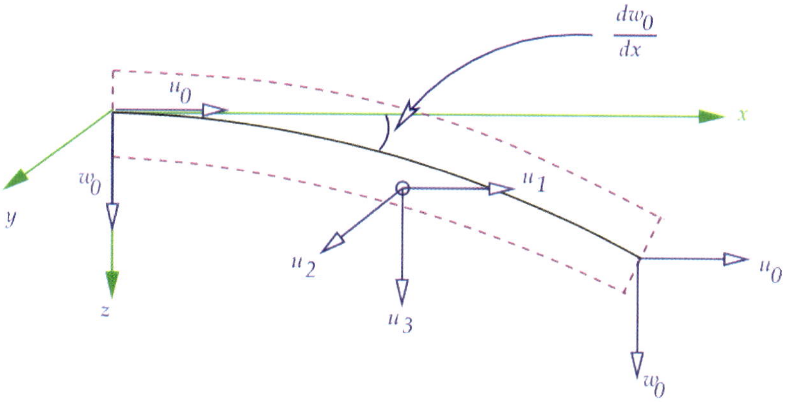

Euler–Bernoulli beam

The Euler–Bernoulli hypotheses that plane sections remain plane and normal to the axis of the beam lead to displacements of the form,

$$v_1 = v_0(x) - z\frac{\mathrm{d}w_0}{\mathrm{d}x} \ ; \ v_2 = 0 \ ; \ v_3 = w_0(x)$$

Using the definition of the Lagrangian Green strain from finite strain theory, we can find the *von Karman strains* for the beam that are valid for large rotations but small strains. These strains have the form,

$$\varepsilon_{11} = \frac{\mathrm{d}u_0}{\mathrm{d}x_1} - x_3\frac{\mathrm{d}^2 w_0}{\mathrm{d}x_1^2} + \frac{1}{2}\left[\left(\frac{\mathrm{d}u_0}{\mathrm{d}x_1} - x_3\frac{\mathrm{d}^2 w_0}{\mathrm{d}x_1^2}\right)^2 + \left(\frac{\mathrm{d}w_0}{\mathrm{d}x_1}\right)^2\right]$$

$$\varepsilon_{22} = 0$$

$$0\varepsilon_{33} = \frac{1}{2}\left(\frac{\mathrm{d}w_0}{\mathrm{d}x_1}\right)^2$$

$$\varepsilon_{23} = 0$$

$$\varepsilon_{31} = \frac{1}{2}\left(\frac{\mathrm{d}w_0}{\mathrm{d}x_1} - \frac{\mathrm{d}w_0}{\mathrm{d}x_1}\right) - \frac{1}{2}\left[\left(\frac{\mathrm{d}u_0}{\mathrm{d}x_1} - x_3\frac{\mathrm{d}^2 w_0}{\mathrm{d}x_1^2}\right)\left(\frac{\mathrm{d}w_0}{\mathrm{d}x_1}\right)\right]$$

$$\varepsilon_{12} = 0$$

From the principle of virtual work, the balance of forces and moments in the beams gives us the equilibrium equations,

$$\frac{\mathrm{d}N_{xx}}{\mathrm{d}x} + f(x) = 0$$

$$\frac{\mathrm{d}^2 M_{xx}}{\mathrm{d}x^2} + q(x) + \frac{\mathrm{d}}{\mathrm{d}x}\left(N_{xx}\frac{\mathrm{d}w_0}{\mathrm{d}x}\right) = 0$$

where $f(x)$ is the axial load, $q(x)$ is the transverse load, and

$$N_{xx} = \int_A \sigma_{xx} \mathrm{d}A \; ; \; M_{xx} = \int_A z\sigma_{xx} \, \mathrm{d}A \, .$$

To close the system of equations we need the constitutive equations that relate stresses to strains (and hence stresses to displacements). For large rotations and small strains these relations are,

$$N_{xx} = A_{xx}\left[\frac{du_0}{dx} + \frac{1}{2}\left(\frac{dw_0}{dx}\right)^2\right] - B_{xx}\frac{d^2 w_0}{dx^2}$$

$$M_{xx} = B_{xx}\left[\frac{du_0}{dx} + \frac{1}{2}\left(\frac{dw_0}{dx}\right)^2\right] - D_{xx}\frac{d^2 w_0}{dx^2}$$

where

$$A_{xx} = \int_A E \, \mathrm{d}A \; ; \; B_{xx} = \int_A zE \, \mathrm{d}A \; ; \; D_{xx} = \int_A z^2 E \, \mathrm{d}A \, .$$

The quantity A_{xx} is the *extensional stiffness*, B_{xx} is the coupled *extensional-bending stiffness*, and D_{xx} is the *bending stiffness*.

For the situation where the beam has a uniform cross-section and no axial load, the governing equation for a large-rotation Euler–Bernoulli beam is,

$$EI\frac{d^4 w}{dx^4} - \frac{3}{2}EA\left(\frac{dw}{dx}\right)^2\left(\frac{d^2 w}{dx^2}\right) = q(x)$$

Beam Deflection

The deformation of a beam is usually expressed in terms of its deflection from its original unloaded position. The deflection is measured from the original neutral surface of the beam to the neutral surface of the deformed beam. The configuration assumed by the deformed neutral surface is known as the elastic curve of the beam.

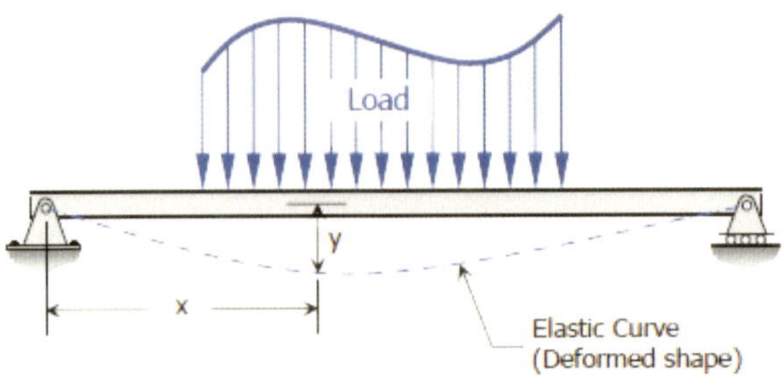

Methods of Determining Beam Deflections

Numerous methods are available for the determination of beam deflections. These methods include:

1. Double-integration method
2. Area-moment method
3. Strain-energy method (Castigliano's Theorem)
4. Conjugate-beam method
5. Method of superposition.

References

- Kenneth L. Kaiser (2004). Electromagnetic Compatibility Handbook. CRC Press. pp. 13–52. ISBN 978-0-8493-2087-3.
- Conductance-82339: universetoday.com, Retrieved 19 June 2018
- Cohen, Robert S.; Spitzer, Jr., Lyman; McR. Routly, Paul (October 1950). "The Electrical Conductivity of an Ionized Gas" (PDF). Physical Review. 80 (2): 230–238. doi:10.1103/PhysRev.80.230.
- Table-of-electrical-resistivity-conductivity-608499: thoughtco.com, Retrieved 27 May 2018
- William D. Greason (1992). Electrostatic discharge in electronics. Research Studies Press. p. 48. ISBN 978-0-86380-136-5. Retrieved 4 December 2011.
- Reactance, technology: britannica.com, Retrieved 17 July 2018
- Rawlins, A. D. (1985). "Note on the Capacitance of Two Closely Separated Spheres". IMA Journal of Applied Mathematics. 34 (1): 119–120. doi:10.1093/imamat/34.1.119.
- Sheet-resistance-theory: ossila.com, Retrieved 26 April 2018
- John D. McDonald (2016). Electric Power Substations Engineering, Second Edition. CRC Press. pp. 363–. ISBN 978-1-4200-0731-2.
- Capacitance, electricity-and-magnetism: tutorvista.com, Retrieved 10 March 2018
- Spitzer, Jr., Lyman; Härm, Richard (March 1953). "Transport Phenomena in a completely ionized gas" (PDF). Physical Review. 89 (5): 977–981. doi:10.1103/PhysRev.89.977.
- 53875-resistors-capacitors-inductors: livescience.com, Retrieved 31 March 2018
- "Joseph Henry". Distinguished Members Gallery, National Academy of Sciences. Archived from the original on 2013-12-13. Retrieved 2006-11-30.
- Stress-strain-diagram, mechanics-and-strength-of-materials: mathalino.com, Retrieved 15 April 2018
- Kim, Seok; Kim, Shin-Ae; Jung, Goeun; Kwon, Kee-Won; Chun, Jung-Hoon, "Design of a reliable broadband I/O employing T-coil", Journal of Semiconductor Technology and Science, vol. 9, iss. 4, pp. 198–204.

4

Sensing and Actuation

A sensor is a device that is designed to detect and measure changes in its environment and relay the data to other electronics. An actuator is the part of machinery that controls the movement or operation of a system or mechanism. The diverse aspects of sensing and actuation, as well as the working principles of electrostatic-sensitive device, electrostatic sensors, electrostatic actuators, thermal actuators, magnetic actuators, etc. have been discussed in detail in this chapter.

Electrostatic-sensitive Device

An electrostatic-sensitive device (ESD) is an electrical component that can be damaged by the build-up of static charges on tools, people or semiconducting or non-conducting materials. Computer parts such as the CPUs, which are densely packed with transistors, are susceptible to such damage. Some common such devices that are electrostatic-sensitive include computer cards, MOSFET transistors, laser diodes, CMOS ICs, TTL chips, LEDs and high precision resistors.

Electrostatic Discharge (ESD)

Electrostatic discharge (ESD) is the release of static electricity when two objects come into contact. Familiar examples of ESD include the shock we receive when we walk across a carpet and touch a metal doorknob and the static electricity we feel after drying clothes in a clothes dryer. A more extreme example of ESD is a lightening bolt. While most ESD events are harmless, it can be an expensive problem in many industrial environments.

ESD first requires a build-up of an electrostatic charge. This occurs when two different materials rub together. One of the materials becomes positively charged; the other becomes negatively charged. The positively-charged material now has an electrostatic charge. When that charge comes into contact with the right material, it is transferred and we have an ESD event. The heat from the ESD event is extremely hot, although we do not feel it when we are shocked. However, when the charge is released onto an electronic device such as an expansion card, the intense heat from the charge can melt or vaporize the tiny parts in the card causing the device to fail. Sometimes an ESD event can damage a device, but it continues to function. This is a called a latent defect, which is hard to detect and significantly shortens the life of the device.

Many electronic devices are susceptible to low voltage ESD events. For example, hard drive components are sensitive to only 10 volts. For this reason, manufacturers of electronic devices incorporate measures to prevent ESD events throughout the manufacturing, testing, shipping, and handling processes. For example, an employee may wear a wrist strap when working with devices or may wear ESD control footwear and work on an ESD floor mat that causes the electrostatic charge to go into the ground instead of into the device. Sensitive devices can be packaged with materials that shield the product from a charge.

Damage from ESD

When a statically-charged person or object touches an electrostatic discharge sensitive (ESDS) device, there is a possibility that the electrostatic charge could be drained through sensitive circuitry in the device. If the electrostatic discharge possesses sufficient energy, damage could occur in the device due to localized overheating. Generally, devices with finer geometries are more susceptible to damage from ESD.

The modes in which ESD damage occurs are:

- Discharge to the device
- Discharge from the device
- Field-induced discharge.

Classifications of ESD Sensitivity

Electrostatic discharge sensitive (ESDS) parts are commonly characterized to three defined models:

- Human Body Model (HBM)
- Machine Model (MM)
- Charged Device Model (CDM)

Based on the models used, the ESDS parts can be classified in accordance with the following table (per MIL-STD-1686C, with HBM subgroups per ESD STM5.1-2001). It should be noted that the HBM, MM and CDM voltage levels do not correlate with each other.

ESD Model	ESD Classification	Voltage Range
Human Body Model (HBM)	0	0V – 249V
	1A	250V – 499V
	1B	500V – 1999V
	1C	1000 – 1999V
	2	2000 – 3999V
	3A	4000 – 7999V
	3B	>= 8000V

Machine Model (MM)	M1	0V – 100V
	M2	101V – 200V
	M3	201V – 400V
	M4	401V – 800V
	M5	>800V
Charged Device Model (CDM)	C1	0V – 124V
	C2	125V – 249V
	C3	250V – 499V
	C4	500V – 999V
	C5	1,000V – 1,499V
	C6	1,500V – 2,999V
	C7	>= 3,000V

Damage from ESD in an Electronic Device

There are basically two categories of damage from ESD:

- *Catastrophic damage* – the electronic device is rendered inoperable immediately after the ESD event. A semiconductor junction or a connecting metallization could have been damaged by the electrostatic discharge.
- *Latent damage* – the electronic device appears to be working fine following the ESD event. However, the sensitive circuitry has been damaged and could fail to operate properly at some time in the future.

Detection of Static Charges in the Work Area

A commonly used tool for the detection of static charges is the electrostatic field meter. This tool, when used in conjunction with regular audits on the production floor, is very effective in detecting the presence and magnitude of static charges. Care should be exercised to ensure adherence to the measurement distance as the meter is calibrated to specific distances from the measuring plane. Please consult the user's guide on the correct measuring distance for the specific electrostatic field meter model that is being used.

Protection for Electrostatic Discharge Sensitive (ESDS) Devices

(A) Work area:

- It is essential to handle ESDS devices at static-safe workstations. This will prevent yield loss (through catastrophic damage) or, worse, potential reliability failures in the field (through latent damage).
- Where it is impractical or impossible to use antistatic wrist-straps or remove items that are composed of insulative materials at a static-safe workstation, use an air ionizer designed to neutralize electrostatic charges or apply

topical antistats to control generation and accumulation of static charges.

- When an air ionizer is utilized, it is vital that maintenance procedures and schedules are adhered to in order to ensure that ions generated by the ionizer are sufficiently balanced.
- Avoid bringing sources of static electricity (as shown in page 1) within 1 meter of a static-safe work bench.
- Where it is necessary to use air-guns, use special models that do not generate static charges in the air stream.

(B) Personnel:

- Any accumulated charge on the body of the human operator should be discharged first before opening the protective container with ESDS devices inside. The discharge can be accomplished by putting a hand on a grounded surface or, ideally, by wearing a grounded antistatic wrist-strap.
- The use of an antistatic smock for each worker is highly recommended.
- Education and training on ESD preventive measures is invaluable.
- A regular audit is also helpful in supporting an ESD program.

(C) Packaging and Transportation:

- ESDS devices should be contained in a static protective bag or container at all times during storage or transportation.

Static-safe Work Bench

The diagram below shows a typical static-safe work bench. The table top is covered by a static dissipative mat which is grounded through a 1 Meg-ohm resistor. This resistor is required in order to protect the users of the static-safe work bench – in the event that the ground becomes electrically live, the resistor will prevent electrical shock at the work bench. The same safety requirement holds true for the antistatic wrist-strap as well.

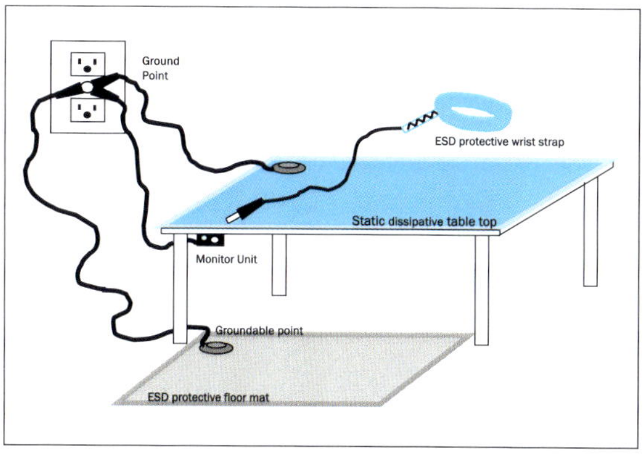

Electrostatic Sensors

Around electrified object, an electric field that is proportional in strength to the amount of charge is produced. Electrostatic sensors detect the intensity of this electric field and calculate it as electric potential.

Detection of Electric Field

When a detection electrode is brought close to an electrified body, an electric charge that is proportional to the intensity of the electric field is induced in the detection electrode due to "electrostatic induction". The electrostatic sensor opens and closes a tuning-fork vibrating plate called a chopper in front of the detection electrode in order to cancel out DC noise and perform higher precision measurement. The sensor detects the intensity of the electric field by receiving the induced electric charge as a communication signal.

Conversion to Electric Potential

Electric potential is proportional to the intensity of the electric field, but the intensity of the electric field gets smaller as it gets further away from an electrified object. Therefore, the electrostatic sensor sets a distance between the electrified object and the sensor using a controller and a corrected calculation of the electric potential is performed.

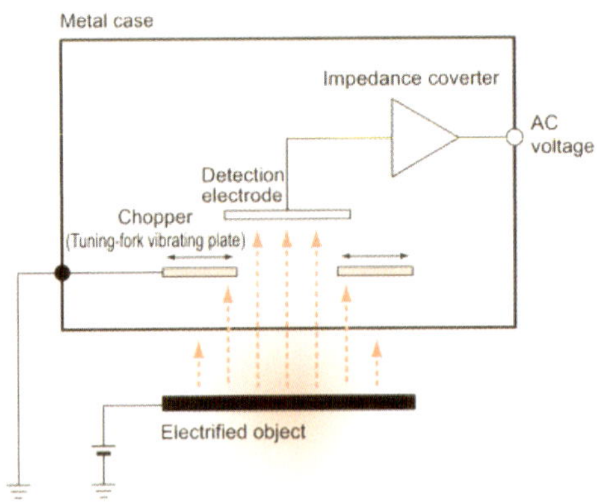

Characteristics based on the principle of electrostatic sensor detection

Since an electric field relies on the measurement distance, you need to fix the sensor at a set distance in order to perform a high precision measurement. The electric field that is produced by an electrified object spreads concentrically out from the electrified object. Therefore, the electrostatic sensor that detects the electric field measures a wider range as the measurement distance increases. Moreover, the existing electrostatic sensors and electrometers have the same range characteristics since they all detect the electric field.

Working of Electrostatic Sensors

When writing is fashioned on a sheet of paper resting upon other pages, the indentations or impressions produced are transferred to those below. These transferred impressions can be detected using an EDD.

In some situations, a questioned document such as a ransom note, or an extortion letter, may exist which can be determined to be the source of indentations detected on another piece of paper (e.g., an offender's notepad). Alternatively, indentations detected on a business contract might match information present on another such document. In some situations this would be an entirely innocent finding; however, if the two businesses are supposed to be operating independent of one another, then the finding could be significant. Decipherable indentations may also provide valuable information even when a second document is not present or cannot be located. For example, an anonymous letter may bear impressions of writing that relate to some mundane activity of the offender which could ultimately lead an investigator to a particular suspect.

The electro-physical basis whereby an EDD actually works is complex. The original theory suggested that the paper sandwiched between the grounded platen and the mylar charging film acted as a type of capacitor with the change in capacitance being due to differing compression of the paper. This led to models like the 'Thickness Variation Theory' and the 'Surface Variation Theory'. However, it turns out that a detectable 'indentation' is not due to the physical pressure applied to the writing instrument as one might expect. Rather, Seward in 1998 and 1999 proposed an alternative theory explaining the detection capability of an EDD as being due to a surface charge effect created by paper-to-paper friction specifically in the area where a writing instrument is pressed down into the top-most sheet of paper. Seward's model was based upon "charge transport through the Mylar-paper-platen structure" and is appropriately called the 'charge transport model'.

Subsequent testing, demonstrated that the charge transport model, while not perfect, is sound. This testing also clarified that areas of indentation are less negatively charged than surrounding areas. It is this relative difference in potential that causes the toner to be attracted to the areas of indentation, rather than other areas on the mylar surface.

Seward's model also helps to explain two unusual phenomena sometimes observed when using an EDD:

- 'Pure' indentations caused by impact printing, for example, may not produce good EDD results
- Indirect secondary 'impressions' may appear that are caused by lateral relative motion between two sheets of paper when the source sheet bears significant embossing.

Additionally, indentations can sometimes not develop very well even though they are clearly visible. This may occur, for example, with very heavy or deep impressions or with impressions in glossy heavy-weight paper such as that used for brochures. In such situations, it is possible to use digital scanning and image-processing to document such writing impressions.

Practical use

Despite the complexity of the underlying theoretical mechanism, the practical use of an EDD is straightforward. Most devices are similar in their operation. The following are the key steps in using an EDD to visualize indentations:

- Evaluation of material
  - Assess adequacy of material for examination: An EDD works best when used on a single sheet of clean, smooth paper (that is, paper without wrinkles, creases or stains). The presence of any of these factors may preclude effective examination or, at the minimum, mean that the results will be limited. Heavy-weight, coated or treated papers often produce poor results. The purported age of the document is not generally a factor of concern. Latent indentations have been developed on documents older than fifty years.
  - Examination using side (oblique) lighting: As a general rule, all items should be examined using side lighting. This is important because visual examination may detect deep indentations even though the EDD may fail to develop those indentations properly.
  - Documents that have been subjected to high levels of humidity will not generally retain indentations. As a result, other forensic examinations involving the wetting of a document, such as fingerprint development using a ninhydrin solution, should not be done until after examination for indentations.
- Preparation
  - Humidification: An EDD works best when the relative humidity is less than 60%. Some studies have also suggested that absolute humidity is important. To help ensure proper humidification, most manufacturers provide a 'humidification chamber' that can be used to prepare a document for examination. It is important to note that over-humidification is also possible so care must be taken to not actually wet the document.
  - Fitness-for-use (FFU) test; AKA, control sample: Since an EDD may fail to detect indentations when they are actually present (for a variety of reasons), it is important to check on the equipment's performance using a control sample (commonly called a FFU test) known to have indentations. Recommendations include using a FFU test sample at the beginning and end of a

series involving the examination of multiple questioned documents, or placing a small FFU test sample adjacent to each questioned document on the platen. The FFU test can be prepared by hand using a similar writing instrument and similar paper. Alternatively, a special device call the 'Gradient®' can be used to produce control indentations that vary in depth in a known manner thereby permitting the examiner to determine not only that the EDD is working but also the degree to which it is sensitive. It is not essential to precisely match the questioned document in terms of paper type or pen type since the objective is to show that the device is functioning properly.

- Placement on platen: The goal is to place the questioned document flat on the platen surface with few wrinkles or distortions (ideally, none). The document should not extend beyond the platen edges and enough room should be present around the document to allow the charging film to seal against the platen surface.
    - Type of document: An EDD works well with a single sheet of paper up to a legal-sized sheet. However, if there are concerns about biological or chemical contamination, due to the condition of the questioned document, it is possible to use a second sheet of clean paper, e.g. copier paper, known to be free of indentations as a barrier between the platen and questioned document. Books and other multipage items can also be examined in some instances but these require additional preparation. Ideally, individual pages should be removed from such items but this is not always possible. As an alternative, a sheet of conductive material (e.g., aluminum foil) can be used to contact the conductive surface.
    - Covering with charging film: The charging film should cover the document completely and contact the platen all the way around the document. The charging film serves two purposes: it protects the document from subsequent toner application and it takes the electrostatic charge required to 'develop' any latent indentations on the document. The film should be laid over the document taking care to avoid wrinkles or excessive stretching of the film. Most EDD devices have a roll of bulk film positioned adjacent to the device for ease in application. The charging film will adhere to the platen due to suction applied through the surface of the platen.
- Charging of surface: The top surface of the charging film must be electrostatically charged. To achieve this most EDDs use a shielded (grounded) handheld unit containing a high-tension corona wire which is turned on and literally waved over the entire surface of platen for several seconds criss-crossing the platen area. The objective is to produce an evenly distributed electrostatic charge in the air above the surface of the charging film. After passing the unit over the platen, the examiner waits for a few minutes until the charge has been passed to the surface of the charging film.

- Application of toner to develop indentations. The EDD is able to visualize indentations because the surface is differentially charged depending upon whether or not an indentation is present. The negatively charged toner particles are attracted to areas where there are indentations in the paper surface. Once an electrostatic charge has been applied to the charging film, the latent indentations can be developed. This is done using a powder toner very much like toner found in any electrophotographic printing device. There are various methods used for this purpose.
    - Cascade method: Some EDDs have a platen that is hinged on one edge to permit it to be tilted. A mixture of toner and small carrier beads (made of glass or similar non-conductive material) is then poured over the surface of the document from one side to the other. The beads, and excess toner, are captured on the lower edge of the device to be reused. This process can be repeated several times in order to develop all areas of the document.
    - Aerosol spray method: Some EDDs provide the ability to spray toner over the surface of the charging film. In these systems a protective shell or chamber is fitted that covers the entire platen. At one end there is a nozzle connected to a reservoir of toner. The examiner then manually activates a pump that sprays toner into the chamber creating a cloud of fine toner that settles down onto the surface of the charging film. Ideally the cloud is randomly distributed but often this method produces a relatively distinctive pattern of toner on the film.
    - Brush method: With some EDDs the toner is applied using a magnetic brush such as would be used for development of latent fingerprints. This is a manual process that is quite tedious and requires meticulous control of the brush. But the method works well for selective development of subtle indentations.
    - Toner pad method: Similar to the brush method described above but uses a special pad with a built-in toner reservoir. The toner pad is placed on the surface of the charging film and moved across the surface by hand.
- Assessment of results. After the development process is complete the examiner must assess the results of the test process.
    - FFU test results: Since the questioned document may or may not have indentations, the correct method to assess the functionality of the EDD is through inspection of the FFU test sample.
        1. If indentations are developed from the FFU test sample, the EDD can be considered as functioning correctly.
        2. If no indentations developed from the FFU test sample, the EDD is not working.

3. If the FFU test was positive and no indentations are developed in the area of the questioned document, then the assessment will be that no detectable indentations have been developed. It is important to note that this does not mean the questioned document was never in contact with other items of interest. The absence of detectable indentations may be due to one or more limitations causing the indentations to not be detected.
4. Note that the presence of indentations in the document area is prima facie evidence that detectable indentations exist such that the FFU test is redundant.

- Preservation of results. It is important to record and/or preserve the results of a test, especially if the test is positive (i.e., indentations were developed). This can be done in at least two ways.
    - Adhesive lift: In this method, a sheet of fixing film is used to preserve the developed indentations. The fixing film is usually a relatively rigid sheet of self-adhesive, transparent polymer film. It is applied by removing the backing material and carefully laying the fixing film directly onto the charging film, thus trapping and fixing the toner in place between the films. Pressure is then carefully applied to the top of the fixing film by hand or with a small roller. The resulting sandwich is referred to simply as a "lift", or EDD lift. Interpretation of the developed indentations is generally done using the lift after removal from the document because this permits the examiner to see the developed indentations on their own without interference from any visible markings on the document.
    - Photography/scanning: The results may also be captured using a camera, either directly from the charging film or from the lift after removal from the document. Scanning is also an option but only when using the lift itself.

Aside from visualizing indentations, the method can be used to detect the presence of fresh fingerprints, fiber disturbance on the paper surface (e.g. relating to mechanical erasure of information), or the presence of footwear impressions on paper.

MEMS Magnetic Field Sensors

MEMS magnetic field sensor is a microelectromechanical systems (MEMS) device, which is used at small-scape for the detection and measurement of magnetic fields. Magnetic field sensors based on MEMS technology, depending on their operation principle and magnetic range, have a great potential for numerous applications in several fields spanning from vehicle detection and control to mineral prospecting and metal detection as well as to non-distructive testing and medical diagnostics.

Resonant Magnetic Sensors

Resonant sensors exploit Lorentz force of resonating micromachined structures. These sensors can detect magnetic fields up 1 mT with a resolution down to 1 nT. Such devices are normally based on MEMS technologies, are small in size (order of millimetres), and promise all the advantages related to the employment of fabrication microtechnologies including multifuntionalities and integration of mechanical and electronic components on a single chip.

Resonant magnetic field sensors use resonant structures that are excited at their resonant frequencies by Lorentz forces. Such devices are able to give an amplified response if excited at frequencies equal to the resonant frequencies or vibrational modes of the structures.

These structures commonly consist of clamped-free beams or clamped-clamped beams or torsion/flexion plates. In figure it is shown a schematic diagram of the Lorentz force principle acting on a clamped-clamped beam resonant structure. This device can be designed for example to resonate to its first resonant frequency, associated to its first flexural vibration mode. This beam, exposed to an excitation source with a frequency equal to its first resonant frequency, will have a maximum deflection at its midpoint. In order to excite the device a metallic loop is placed on the clamped-clamped beam surface where an excitation current (I) flows inside it with a frequency equal to the first resonance frequency. When the beam is exposed to an external magnetic field (B_x) in the x-direction, then a Lorentz force (F_L) is generated.

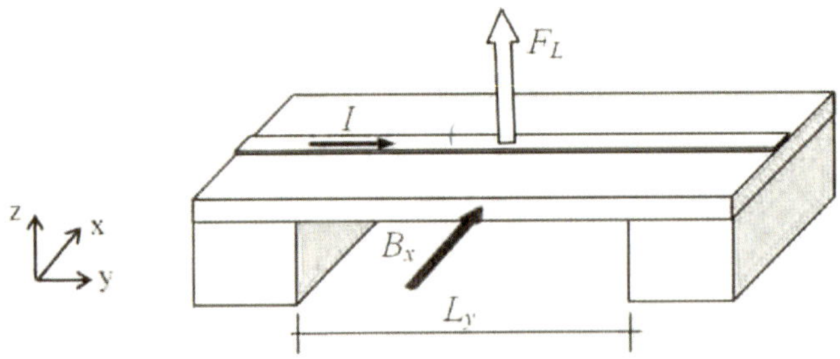

Figure: Schematic diagram of the Lorentz force principle acting on a clamped-clamped beam

This force can be determined as:

$$F_L = lB_x L_y$$

where the flowing current can be expressed as:

$$I = \sqrt{2}1_{rms} \sin(2\pi\ ft)$$

where L_y is the length of the metallic loop perpendicular to the magnetic field, I_{rms} is the root mean square of the current I, f is the frequency and t is the time.

The Lorentz force acts as an excitation source on the clamped-clamped beam, causing an amplified deflection on the midpoint. Thus, the magnitude of the beam deflection depends on the Lorentz force amplitude, which is directly proportional to I and B_x.

The application of an external magnetic field alters deflection/torsion of resonating structures with different shapes that is detected by exploiting different readout techniques.

In fact, such deflections/torsions result in strain which is related to the elastic modulus of the structure material, to the geometrical characteristics of the resonating structures and to the quality factor.

The quality factor is an important parameter of the resonant structures. It defines the bandwidth of the resonator relatively to its central resonant frequency or equivalently it expresses the maximum amplitude of the bending structure taking into account the different damping sources. High quality factors involve better device performance, better resolution and improved insensitivity to the disturbances.

Another parameter of interest in resonant structures is the resonance frequency. Its determination can be obtained by using both analytical models and simulation tools and depends on elastic modulus, density, deflection and geometrical features of the resonant structure.

Moreover resonance frequencies are affected by residual stresses on the structure. For example thermal stresses inside resonant devices causes strains in the structures which in turn involve a shift of the resonant frequency of the structures.

Such sensors are typically fabricated in silicon and polysilicon and main disadvantage of this technology is the resonance frequency shift due to temperature changes and environmental pressure which requires compensation electronic circuits and packaging under vacuum respectively.

To detect deflection of resonant structures different readout techniques have been used including the employment of piezoresistive, optical or capacitive techniques.

Piezoresistive sensing exploits changes in the resistance of piezoresistive elements placed in the hinges of the resonant structure, to detect changes in the output voltage signal as effect of strains originating from motions of beams or plates due to the Lorentz force.

Herrera-May reported on a magnetic field microsensor based on a silicon resonant microplate ($400 \times 150 \times 15\ \mu m^3$) and four bending microbeams ($130 \times 12 \times 15\ \mu m^3$).

The fabrication process is based on bulk micromachining technology on (100) 4” silicon-oninsulator (SOI) wafers. The process starts by growing a thin thermal oxide layer and depositing a silicon nitride layer on a SOI n-type substrate. The nitride layer is removed from the front side of the wafer and is patterned on the backside. Using a second mask, boron is implanted to create four p-type piezoresistors. A 1 μmthick oxide layer is then grown and patterned. The area contacts (120 × 120 μm²) are opened and then an aluminum layer is deposited and patterned to define metallic lines and pads. At this time the silicon substrate is etched from the backside using KOH that stops at the SOI buried oxide, which is then removed.

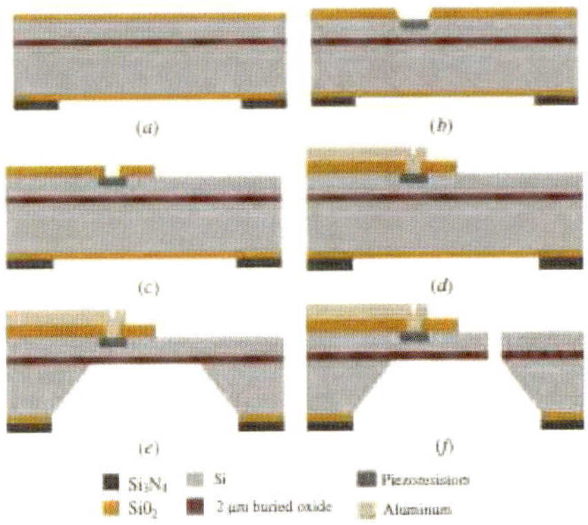

Fig. Schematic diagram of the fabrication process of a piezoresistive resonant magnetic sensor

Finally, the SOI layer is etched by reactive ion etching to define the plate-beam structure.

Figure shows a schematic design of the resonant magnetic field microsensor with an highlight on the plate-beam structure and its working principle.

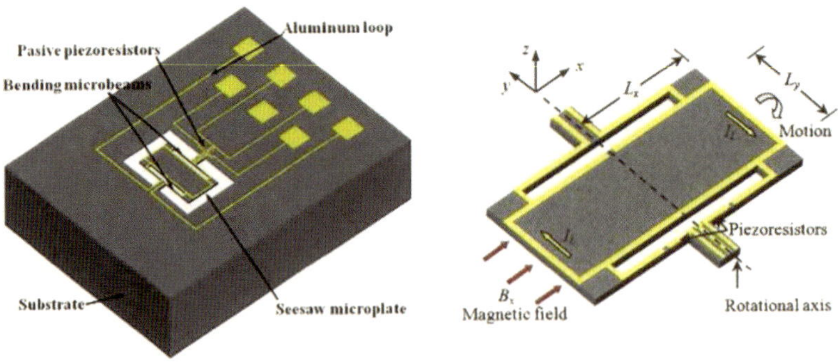

Fig. Schematic design of resonant magnetic field microsensor (left) and highlight on the plate-beam structure and its working principle (right)

One of the main elements of this sensor is the aluminum rectangular loop deposited on the silicon plate. The Lorentz force causes a seesaw motion on the microplate and the bending of microbeams. Four piezoresistors (p-type) are connected in a Wheatstone bridge and two of these are active piezoresistors located on the microbeams. The Lorentz force originates a longitudinal strain on the two active piezoresistors changing their resistance. The change in the resistance of the active piezoresistors produces an output voltage shift of the Wheatstone bridge.

This sensor has a resonant frequency of 136.52 kHz, a quality factor of 842 at ambient pressure, a sensitivity of 0.403 $\mu V \mu T^{-1}$, a resolution of 143 nT with a frequency variation of 1 Hz, and power consumption below 10 mW. However, the sensor registered an offset and linearity problems in the low magnetic field range.

Tapia reported on a piezoresistive resonant magnetic microsensor with seesaw rectangular loop of beams reinforced with transversal and longitudinal beams. This device was designed to be compact and to have high resolution for neurobiological applications. Characteristics of this microsensor are a resonant frequency of 13.87 kHz with a quality factor of 93, a resolution of 80 nT, a sensitivity of 1.2 VT^{-1} and a power consumption of 2.05 mW at ambient pressure. This sensor requires a simple signal processing circuit.

Other examples of piezoresistive magnetic sensors on the microscale have been reported in the literature.

Among the resonant magnetic sensors, there are some of them exploiting the optical detection.

Keplinger reported a resonant magnetic field sensor using U-shaped silicon microbeams and an optical readout system. Figure shows a schematic sketch of the device reported by these authors.

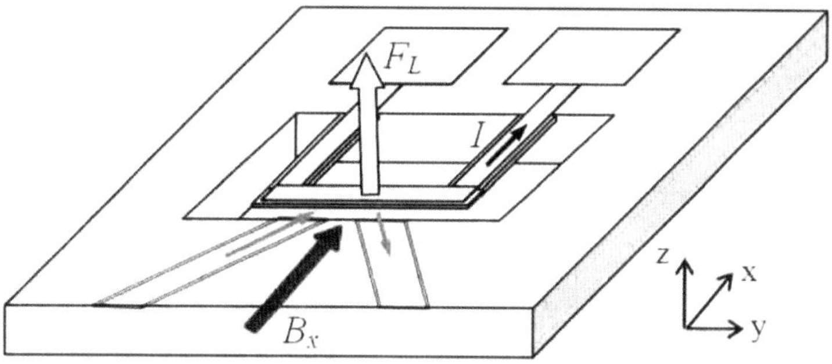

Figure: Schematic sketch of the resonant magnetic sensor with optical readout reported by Keplinger

Devices are processed by both surface micromachining and bulk micromachining techniques. The etching process has been also used to form the groove for the optical fibers. The microbeams contain a gold loop with a thickness of 500 nm. A magnetic field and an ac electrical current generate a Lorentz force, which bends the microbeams. These deflections are measured by optical sensing using an arrangement consisting of two fibers to avoid interference of reflected light. Figure shows a design, in which the emitted light beam is reflected only once at the microbeam front side. This sensor can measure magnetic fields from 10 mT to 50 T at moderate excitation amplitudes. It can be used in harsh environments under mechanical vibrations and low temperatures. The device has a resonant frequency around 5 kHz, a resolution of 10 mT, and a power consumption of few milliwatts. The device needs high current magnitudes thus increasing the temperature and deformation at the silicon microbeam, with a possible resonant frequency shift.

Another example of magnetic sensor based on MEMS technology and having an optical readout has been reported by Wickenden.

Ren reported on a resonant device exploiting the capacitive readout. The magnetic field sensor has been fabricated by using conventional MEMS technology and silicon-to-glass anodic bonding process. The device consists of a low-resistivity silicon plate suspended over a glass substrate by two torsional beams, as shown in figure. This silicon plate acts as electrode of sensing capacitances. Au capacitance plates are fabricated on the glass substrate and a multi-turn coil (Cr and Au layers) is deposited on silicon-plate surface. The Lorentz force causes an oscillating motion of silicon plate around the torsional beams, which produces a capacitance shift between the Au electrodes and the silicon plate. A capacitance detection circuit measured the capacitance change that depends on the magnitude and the direction of the external magnetic field. This sensor required a vacuum packaging to increase its performance. For a pressure of 10 Pa and 150 mV driving voltage amplitude, the microsensor has a resolution of 30 nT in the linear range from 3 μT to 30 μT, a sensitivity of 481 mVT^{-1}, a resonant frequency close to 1380 Hz, and a quality factor around 2500. Nevertheless, it presented a non-linear response from 0 to 3 μT.

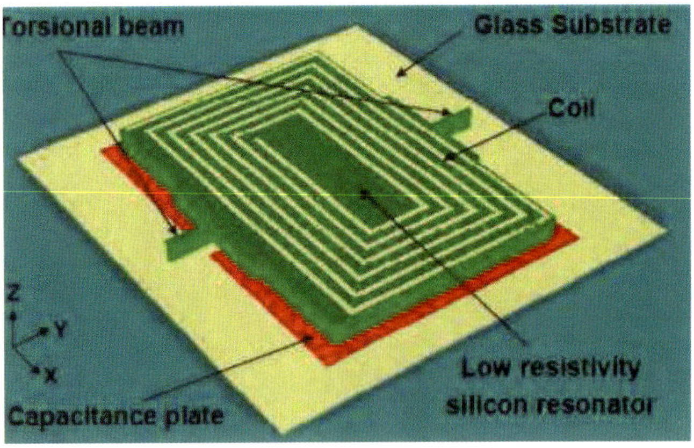

Figure: Schematic design of the resonant magnetic sensor exploiting the capacitive sensing).

Brugger reported on a complex magnetic field sensor with a size of 7.5 mm × 3.2 mm, consisting of an electrostatically driven silicon resonator characterized by interdigitated combs for electrostatic excitation and capacitive detection, an amorphous magnetic concentrator and a pair of planar coils. It requires a complex fabrication process combining MEMS technology based on a silicon-on-insulator (SOI) substrate, the epoxy-resin-based attachment of a thin amorphous magnetic ribbon structured by wet chemical etching, micropatterning of the magnetic concentrator by UV^{-1}aser and vacuum packaging.. For a coil current of ±120 mA, the device offers a sensitivity of 1.91 $MHzT^{-1}$ and a resolution of 1.3 μT. Under a pressure of 10-5 mbar, this microsensor presents a sensitivity of 1 $MHzT^{-1}$, a resolution of 400 nT, and a quality factor around 2400. It does not need a complex feedback and modulation electronics.

J. Kyynäräinen reported on resonant micromechanical magnetometers based on capacitive detection for 3D electronic compasses. The sensors has been fabricated by exploiting aligned direct bonding of a double side polished silicon wafer and a SOI wafer. Devices operated in vacuum to reach high enough Q values. Magnetometers measuring the field component along the chip surface have a flux density resolution of about 10 nT/√Hz at a coil current of 100 μA. Magnetometers measuring the field component perpendicular to the chip surface are currently less sensitive with a flux density resolution of about 70 nT/√Hz.

Fluxgate Sensors

Fluxgate sensors are inductively working sensors composed of excitation and sensing coils around a ferromagnetic core for detecting static and low frequency fields.

Fluxgate sensors can detect magnetic fields up to approximately 1 mT with a maximum resolution of 100 pT. Classical fluxgate sensors are expensive and have a big size. However in recent years a great effort was devoted to manufacturing micro fluxgate sensors using microfabrication technologies. Beside the small size, the advantages of micro fluxgate sensors are small weight, low power consumption, low cost in mass production and the possibility of on-chip electronics integration. The principal disadvantage is related to the fluxgate sensor parameters dramatically degrading when reducing the core size, which leads to low sensitivity and high noise level.

The fluxgate operation principle can be illustrated with the simple layout in figure. A ferromagnetic core immersed in an external magnetic field B_x is surrounded by an excitation coil which provides an ac excitation current I. This current periodically saturates the soft magnetic material of the sensor core at a frequency twice the excitation frequency.

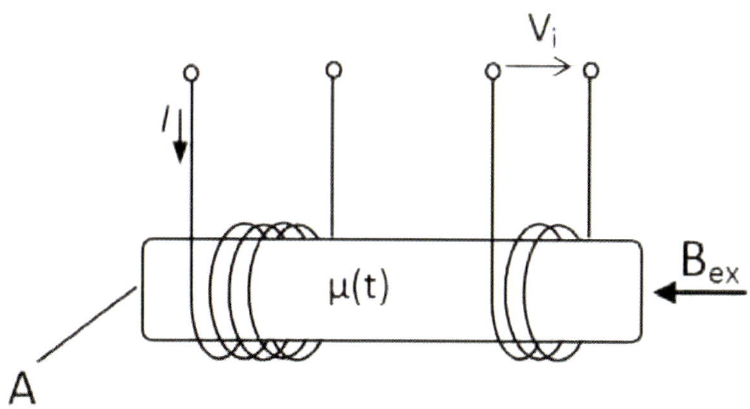

Fig. The basic fluxgate principle.

A voltage in the pick-up coil (Vi) is generated due to the changing magnetic flux (Φ). From the Faraday's law:

$$V_i = \frac{d\Phi}{dt} = \frac{d\left(NA\mu_0\mu_r(t)H(t)\right)}{dt}$$

where μr is the relative permeability, μ_0 is the vacuum permeability, N is the number of turns and A is the cross sectional area, that we consider here as a constant. By expanding, the equation above becomes:

$$V_i = \frac{NA\mu_0\mu_r dH(t)}{dt} + \frac{NA\mu_0 H d\mu_r(t)}{dt}$$

where H is the field in the core material and is lower than the measured field H_{ex} in the open air due to demagnetization:

$$H = H_{ex} - DM$$

where D is the demagnetising factor, $H_{ex} = B_x/\mu o$ and M is the magnetization.

The first term in the equation above is the basic induction effect, and causes interference. Fluxgate operation is based on the second term, due to the variation of the core permeability with the excitation field. By considering the effect of demagnetization, the basic fluxgate equation becomes:

$$V_i = NA\mu_0\mu_r H_{ex} \frac{1-D}{\left(1+D\left(\mu_r(t)-1\right)\right)^2} \frac{d\mu_r(t)}{dt}$$

The output voltage is on the second harmonics of the excitation frequency, as permeability reaches its minimum and maximum twice in each excitation cycle.

In accordance with the shape of the magnetic core, parallel-type fluxgate sensors fall into the categories of single core, dual core, ring-type core, racetrack type core. The configuration of figure is single core type. In order to eliminate the induction effect, a dual core configuration has been proposed, as showed in figure.

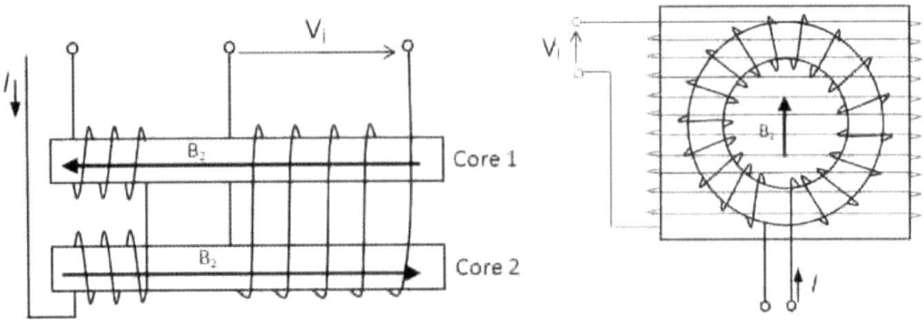

Fig. Dual core (left) and ring type (right) configurations of a fluxgate sensor.

The driving coil is wound in opposite direction around the two cores, thus the induced magnetization fields are opposite in sign. If no external field is applied, the voltage induced in the sensing coil is zero in the ideal case. When an external field is present, a voltage is induced due to the differential change of the permeability.

High sensitivity can be achieved by increasing the number of turns N (if N is very high coil parasitic capacitance limits the sensitivity), by decreasing the demagnetization factor D or by increasing the excitation frequency, because (dHex/dt) ~ f up to frequency values that make eddy currents negligible.

Such devices can be also classified in parallel type and orthogonal type fluxgate sensors depending on the excitation field is parallel or perpendicular to the sensitive axis of the sensor.

Typically these devices contain solenoid systems wiring magnetic cores consisting of a permalloy or an amorphous material.

Figure shows also the ring type configuration. The closed geometry of the ring core has a lower sensitivity but better noise performance, due to the absence of open ends.

A closed core made with oval geometry (race-track fluxgate sensor) lead to a lower demagnetization factor, then to higher sensitivity and less sensitivity to perpendicular fields.

In the orthogonal fluxgate the excitation coil is absent, and the sensor is excited directly by the current flowing through the core.

Three basic types of miniature fluxgate can be distinguished (Ripka & Janosěk, 2010): plane type sensors with flat coils, PCB-based devices with solenoids made by tracks and vias and 3D type sensors with micro solenoids.

While plane type sensors are typically fabricated by standard CMOS processes, MEMS microfluxgate sensors exploit advanced microfabrication technologies to realize three-dimensional coils or three-dimensional cores.

One of the first work which proposed MEMS technology for the development of fluxgate sensors has been reported by Liakopoulos & Ahn. In this work the authors presented a micro-fluxgate magnetic sensor based on micromachined toroidal type planar coils. In this fluxgate sensor a rectangular-ring shaped magnetic core has been chosen. The operation principle is based on the second harmonic. Excitation and sensing coils as well as permalloy magnetic cores were fabricated by a UV-LIGA thick photoresist process and electroplating techniques to realize a planar three-dimensional magnetic fluxgate sensor on silicon wafers. Excellent linear response over the range of -500 mT to +500 mT, with a sensitivity of 8360 VT-1 and a resolution of 60 nT was achieved with this device. The total response range of the sensor is -1.3 to +1.3 mT. The power consumption is around 100 mW for an operational frequency range of 1-100 kHz.

Woytasik proposed an alternative fabrication process based on copper micromoulding to realize planar microcoils and microsolenoids for MEMS based magnetic sensors on flexible substrate. Figure shows main steps for solenoid fabrication process.

The main steps for solenoids fabrication consist of the realization of the bottom conductor lines and of the air bridge by copper electrodeposition overflow. The second exposure process uses a gray-tone mask to vary spatially the exposure dose deposited into the photoresist and then to modulate the remaining photoresist thickness after development. The process ends with mould removal.

This technology has been employed for the realization of a micromachined fluxgate sensor.

The sensor, schematically shown in figure, consists of a 30 μm thick electroplated permalloy core, with 56 excitation turns giving a total resistance of 2 Ω and 11 sensing turns. A sensitivity of 650 VT-1 was achieved for a 5.5-mm-long sensor with 14 mW power consumption. The noise is 32 nT/√Hz @1Hz, and the practical resolution is 1 μT.

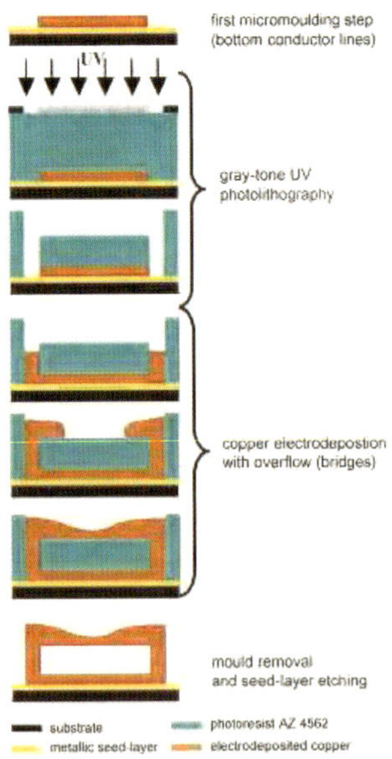

Fig. Main steps of solenoid fabrication process

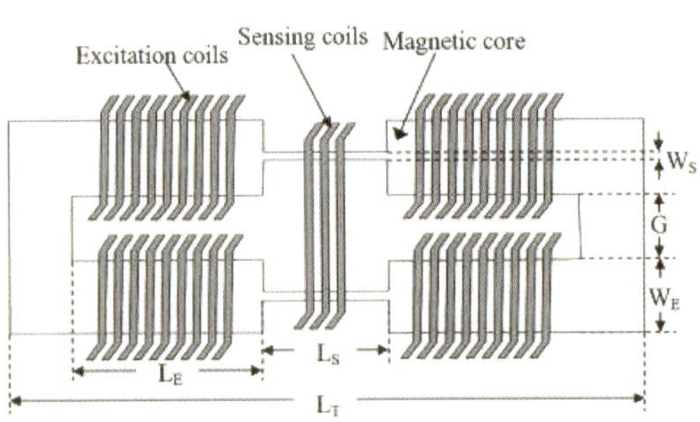

Fig. Schematic view of the MEMS based fluxgate sensor

Other Examples of MEMS-based Fluxgate Devices

Most of the microfluxgate exploiting MEMS technologies are parallel sensors.

A miniature orthogonal fluxgate realized by exploiting MEMS technologies with a planar structure formed by a permalloy layer electrodeposited on a rectangular copper conductor has been reported by Zorlu. The sensor structure is reported in figure. The sensor core is only 1 mm in length and the sensor has two flat 60 turn pickup coils.

The overall dimension of the sensor chip is 1.8 × 0.8 mm, the sensitivity is 0.51 mVmT-1 in a linear operating range of ±200 μT. The noise was 95nT/√Hz@1Hz with an excitation power consumption of 8.1 mW.

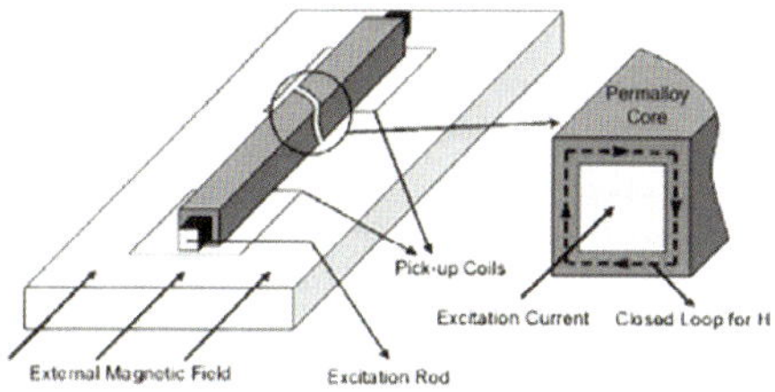

Figure: Sensor structure the orthogonal microfluxgate

Hall Sensors

Hall sensors exploits Hall effect as trasduction principle to detect magnetic field. They are commonly fabricated by standard Complementary Metal-Oxide Semiconductor (CMOS) technology. In general, they are applicable in a range from 1 μT to 1 T and have a die size less than one millimeter.

These sensors can measure either constant or varying magnetic field. The frequency limit is around 1 MHz and operate well in a wide temperature range.

The Hall effect is based on the Lorentz force felt by charge carriers moving in a magnetic field. Figure shows the schematic of the classical configuration where a thin slab of a conductor is placed in a magnetic field Bz. When a current flows in the x direction, the Lorenz force acts in the y direction, determining a charge distribution that counterbalances the force. Therefore a (Hall) voltage VH builds up, as shown in figure.

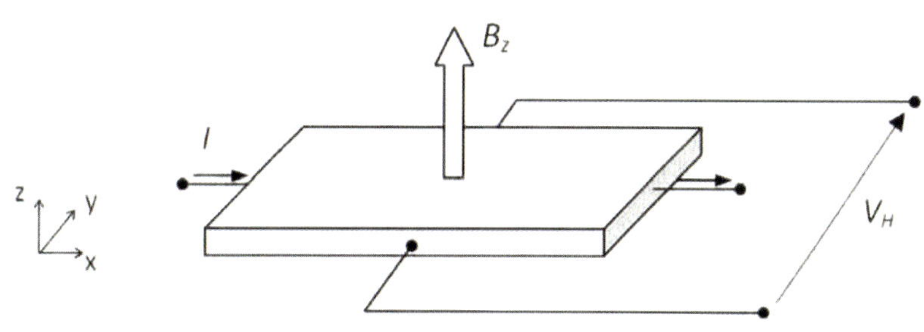

Fig. Hall effect principle.

In the case of a constant current drive, the Hall voltage is given by:

$$V_H = V_{OFF} \frac{B_2 I}{e n_{2D}}$$

where V_{OFF} is the Hall voltage at zero magnetic field (offset voltage) and n_{2D} is the sheet charge concentration given by the product of the bulk charge concentration n and the thickness of the slab t:

$$n_{2D} = nt$$

If the current is constant, lower charge carriers concentration involves higher carriers speed resulting in an higher Lorentz force. Therefore low charge carrier concentration leads to high Hall voltage, thus justifying the extensive employment of semiconductor materials for Hall magnetic sensors with respect to metals.

The current related sensitivity is a key figure of merit of a Hall sensor and can be defined as:

$$S_t = \left| \frac{1 \partial V_H}{1 \partial B_z} \right|$$

As the Hall voltage is related only to the z-axis magnetic field component, Hall magnetic sensors are basically uniaxial devices. Silicon based Hall sensors are widely employed, due to the suitability of integration with electronics. However, magnetic sensors based on silicon may have intrinsic limits to their sensitivity and resolution, which may limit future performance gains. In addition, they need temperature compensation circuits that can include temperature sensor and operational amplifiers (op-amps).

MEMS technology has been employed in this class of devices in order to solve some of their limitations and to find alternatives to silicon as structural material. For example polymer- based devices are interesting alternatives to silicon, particularly when the polymer materials can be functionalized for enhanced specific material properties (e.g, optical, electrical, and mechanical). Mouaziz proposed the realization of SU-8 cantilevers with an integrated Hall-probe for advanced scanning probe sensing applications. To this purpose an innovative release method of polymer cantilevers with embedded integrated metal electrodes has been employed. Figure shows the device fabrication process. On the silicon wafer with 0.5 µm of thermal oxide a 2 µm-thick polysilicon sacrificial layer is deposited. The electrodes of the Hall probe and the metallic thin film electrical connections are obtained by lithographic patterning, metal deposition and liftoff. The device structure is obtained by lithographic patterning of two layers of SU-8 polymer: a 10µm-thick photo-structured layer for the cantilever, and a 200µm-thick layer for the chip body. The releasing method is based on dry etching of a 2µm-thick sacrificial polysilicon layer. A device sensitivity of 0.05 V/AT was achieved together with a minimum detectable magnetic flux density of 9 μT/√Hz at frequencies above 1 kHz at room temperature.

Sunier reported on a vertical Hall sensor with precisely defined active area fabricated by a process combining deep-RIE silicon trench etching and anisotropic TMAH silicon wet etching. The fabrication process is showed in figure. The main process steps are the thermal oxidation of the p-type silicon wafer, the definition of the trench by dry etching, the sidewall implantation and oxidation, another dry etching process to deepen the trenches, TMAH wet etching to release the bottom of the active area, trench oxidation, polysilicon refill and blanket etch, contact formation and metallization and finally XeF2 etching of the polysilicon in the tranches.

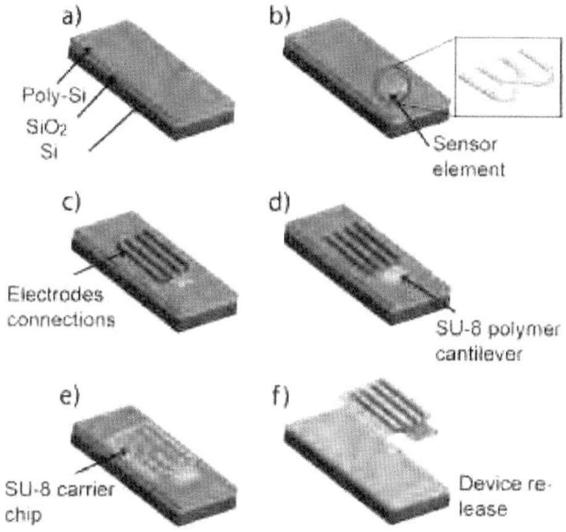

Fig. Schematic illustration of the process for SU-8 cantilever with integrated electrodes

Beside a very high current related sensitivity of up to 1000 V/AT, the improved insulation from the substrate resulted in a more efficient offset compensation and then in a reduced residual offset.

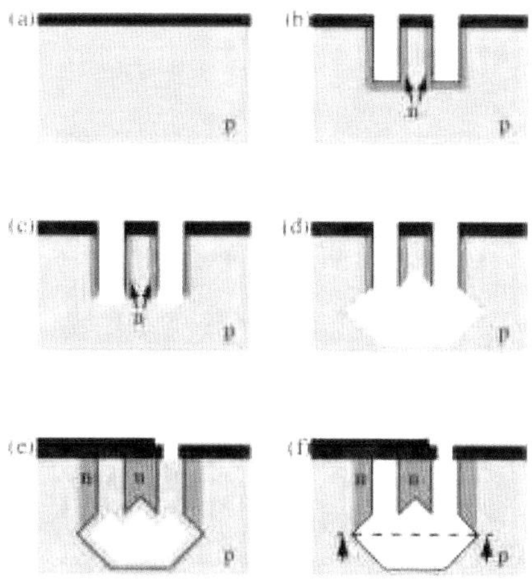

Fig. Simplified fabrication process for trench Hall devices

Estrada proposed a three axis Hall sensor based on MEMS micromachining of SOI-wafers. Three Hall sensors embedded in a flexible polyimide carrier was obtained so that appropriate folding of the structure resulted in three Hall-probes positioned to form an orthogonally-oriented array on three faces of a millimeter-sized cube. The key fabrication process steps are as follow: (a) SOI-wafer; (b) micromachining of the active layer using either wet (TMAH) or dry (DRIE) etching; (e) patterning the Al-film needed for ohmic contacts; (f) deposition and curing of the polyimide film; g) etching of the handle wafer to reach the buried oxide film; (h) final metallization of the contact pads for possible soldering or wire bonding (figure).

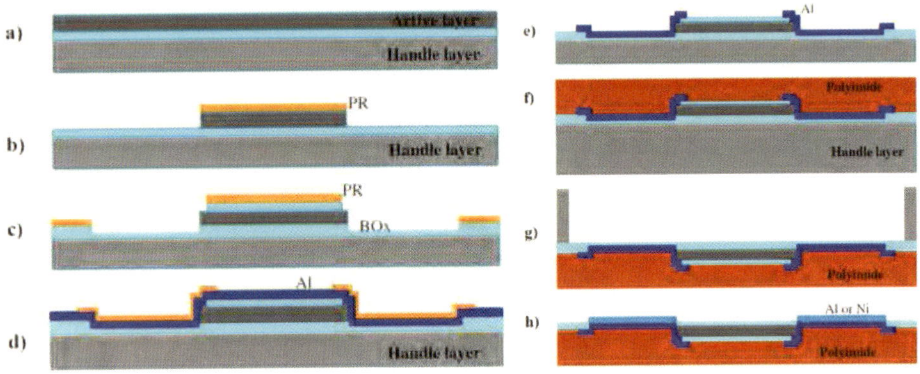

Fig. Key steps for the fabrication of the three-dimensional axis Hall sensor on a flexible substrate (polyimide film)

This 3D-sensor configuration allows vector magnetic field measurements where the advantage is that all its elements have the same magnetic sensitivity of about 100 V/AT.

An integrated three axis Hall sensor based on III-V technology was fabricated by employing a micromachining technique for realizing self positioned structures. The MEMS technique was applied to a GaAs-based heterostructure containing a sensing layer and a strained layer. The selective removal of a third sacrificial layer allows for the relaxation of the strained layer and the self positioning of the sensing part, which has been already processed to realize a Hall sensor element.

The main fabrication steps are as follows: a) Epitaxial growth of a multilayer with sacrificial layer, strained layer, and sensor multilayer; (b) Photolithography and wet etching to define the mesa active region; (c) Photolithography and wet etching to define the hinge region; (d) metallization by lift off (GeAu/Ni/Au); (e) Photolithography and chemical etching to expose the edge of the sacrificial layer; (f) selective etching of the sacrificial layer and self- positioning of the structure.

Current related sensitivity of more than 1000 V/AT both for in-plane and for out-of-plane Hall sensors, demonstrates the effectiveness of this method for realizing fully integrated miniaturized high sensitivity three axis magnetic sensors.

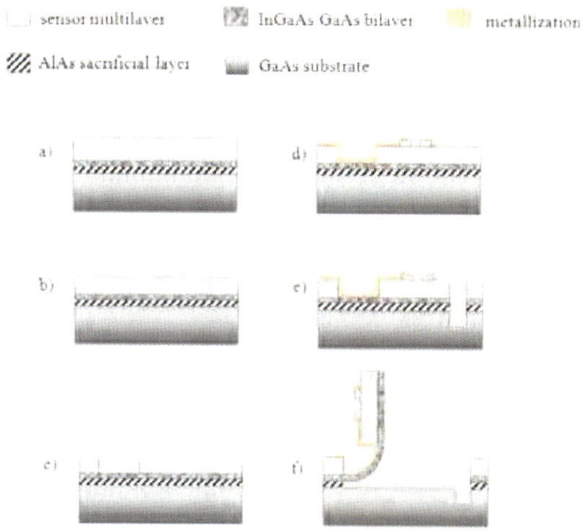

Figure: Three-axis Hall sensor fabrication steps

Figure. shows images of the realized three dimensional magnetic Hall sensor highlighting the out-of-plane sensor.

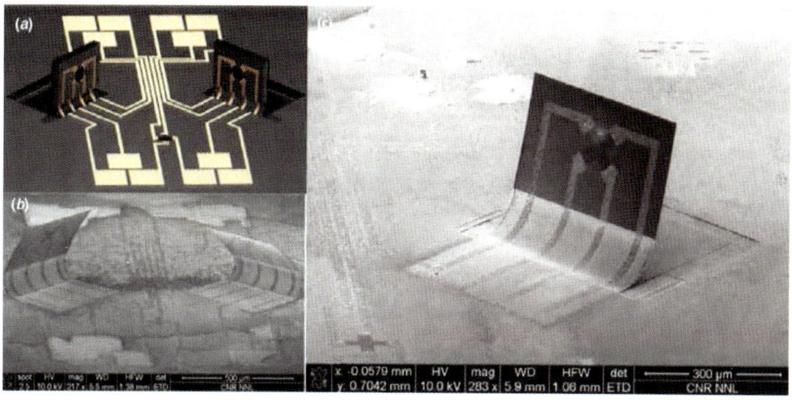

Figure: Fully integrated three-axis Hall sensor a) a schematic of the three-axis Hall sensor; b) Image of a fabricated three axis Hall sensor; c) highlight of an out-of-plane Hall sensor acquired by scanning electron microscopy.

Applications

Magnetic field sensors based on MEMS technology have potential advantages with respect to conventional magnetic field sensors such as small size, light weight, compactness, lower power consumption. Additionally the MEMS technology achieves low-cost sensors by means of batch fabrication techniques and their potential integration with integrated circuits (IC) on a same substrate.

Conventional magnetic sensor classes presented in this paper have different application fields depending on their sensitivity range and minimum detectable magnetic field.

Resonant sensors have a magnetic range up to 1 T with a maximum resolution of 1nT, fluxgate sensors range spans from 100 pT to 1 mT, while Hall sensors have a sensitivity ranging from 1 μT to 1 T. Resonant sensors have lower resolution compared to fluxgate, however they present a wide sensitivity range and they could compete with fluxgate sensors into numerous applications for measuring magnetic fields.

Among the sensors, Hall sensors are the less sensitive devices. Their robustness and simple fabrication process justify their use in hundreds of applications.

Improvements in the microfabrication technologies combined with the employment of new and more performing materials as well as novel design solutions for devices on the microscale could enhance further the resolution, making them suitable for applications requiring very high sensitivity, such as in biomedical field and for the realization of new class of hand-held equipments. On the other hand this technology could help in new solutions for devices in applications requiring low sensitivity.

Magnetic field sensors on the microscale with moderate sensitivity, could be used for vehicle detection and recognition. In fact vehicles moving over ground can generate a succession of impacts on the earth's magnetic field, that can be detected by means of magnetic perturbation using a magnetic sensor, and automatically recognize them by advanced signal processing and recognition method. In this context such sensors could be used for the measure of the speed and size of vehicles for traffic surveillance. Additionally magnetic field sensors can be used in systems containing accelerometers, gyroscopes and pressure devices for vehicle control applications. For example they can be employed in electronic stability program (ESP) systems to help vehicles to be dynamically stable in critical situations like hard braking and slippery surfaces.

Such microsensors can be employed in electronic compasses for sensing earth's magnetic field for GPS systems in order to provide more precise and instantaneous headings to aid navigation for air, ground and underwater systems. Additionally such devices can be used for global positioning systems (GPS) in cell phones due to the requirements of reduced size, low cost and low power consumption.

These magnetic field sensors find employment for the detection of compact ferrous objects. Such objects are of major concern in a number of applications. In environment science there is the need for portable sensors for mineral prospecting like measurements of magnetic properties of rocks, as well as detection of pipeline corrosion where geological ore inclusions generate typical peak magnetic induction in the range 1- 1000 nT. In the military field such devices can be used in systems for the detection and mapping of hidden or unexplosed ordnance (mines, bombs, and artillery shells which have a peak magnetic induction in 10-1000 nT range) as well as for detection of armored vehicles (10000 nT) or submarines (1-10 nT). The performance of these systems can be enhanced by using two or three dimensional array of sensors. This could give additional informations on the size and the depth of the buried objects.

Another application of these sensors is in non-distructive testing for a variety of evaluations including medical implants and aircraft structures, the detection of cracks and corrosion in metals. Archeology is another field requiring systems including magnetic field sensors to resolve non-invasively details, the wide range of artifacts (1-1000 nT magnetic induction range) and cultural objects. This field requires also new means of mapping prehistoric and historic sites in three dimensions rather than traditional two-dimensional methods.

High sensitivity and high resolution magnetic sensors are needed in systems for medical diagnostics. Microfluxgate sensors based on MEMS technology can be employed to build cheap and portable systems for locating metallic foreign objects in the human body.

Ripka, showed that fluxgate sensors can be used for mapping the distribution of ferromagnetic particles in the lungs after they are magnetized by strong DC field. Medical applications requiring precise miniaturized magnetic sensors include tracking devices and systems for monitoring magnetic markers such as magnetic "biscuits" and microbeads. Magnetic biscuits can be used for functional tests of digestive tract, while microbeads are used as markers in biotechnology. New types of fluxgate microsensors are being developed for these applications. Also Hall magnetic sensors have been employed to visualize a magnetically market diagnostic capsule in real time inside human body. Tracking devices using fluxgate sensors can be used for monitoring the 3-D position and also orientation of a small permanent magnet which can be attached to body or medical instrument (such as catheter). Another configuration is being used for tracking the motion of the body at further distances: signals from sensors attached to the body are collected and processed.

Typically Hall magnetic devices, due to their low sensitivity are employed for position sensing, current sensing, speed detection, electronic compasses. Silicon-based Hall sensors are widely employed, due to the suitability of integration with electronics. However, higher sensitivity sensors can be obtained with III-V technology, allowing for applications such as biomolecular function detection. Also recently, the Scanning Hall probe microscopy (SHPM) has been developed based on III-V Hall sensors, allowing for quantitative mapping of nanoscale superconducting and ferromagnetic materials.

Others nowadays applications such as geomagnetic measurements, environmental disturbance measurements as well as navigation systems demand for Hall miniaturized devices capable of measuring the vector magnetic field. Beside the cumbersome solution of mounting three Hall sensors with their sensitive axis orthogonal to each other, integrated three axis devices have been developed by employing silicon technology and the so called vertical Hall effect. However these are often characterized by either different sensitivity for each component of the magnetic field, or by a cross-sensitivity among the direction-components. Furthermore, offset compensation of vertical Hall element is more difficult than in the case of lateral (planar) Hall elements. In this context new materials and device configurations could open the way to realize reliable vector magnetic field sensors to be applied in different fields.

Electrostatic Actuators

Electrostatic actuators rely on the force between two conducting electrodes when a voltage is applied between them. Depending on the arrangement of the electrodes, various types of actuators are possible. However, whatever the electrode arrangement, if there exists mutual capacitance, there will also be an attractive force when a voltage is applied between the electrodes. This force will nearly always be attractive.

Electrostatic forces are a favourite for surface-micromachined structures. This sterns in part from ease of implementation and in part from favourable scaling laws for electrostatic forces. Electrostatic forces are surface forces, hence their favourable scaling. Further, extremely large electric fields can be created with modest voltage because of the small distances involved.

Electrostatic actuators are easy to implement in surface micromachining since structural materials are often conducting. Paired with small manufacturable gaps between components, large electric fields are easily realizable.

One of the main disadvantages of electrostatic actuators is the large voltages necessary. Voltages of several hundred volts are not uncommon. This requires special high-voltage electronics for control, and is a barrier to integration with standard electronics. Fortunately, miniature devices can handle these devices without dielectric breakdown.

Electrostatic actuators can be roughly categorized as linear or rotary. Many types of the linear actuators will be covered in this chapter. The theory behind many of the rotary actuators is similar to their corresponding linear forms.

The main classes of rotary electrostatic actuators are :

- top-drive,
- side-drive, and
- wobble harmonic.

These devices are often called micromotors.

Hydraulic

A hydraulic actuator consists of cylinder or fluid motor that uses hydraulic power to facilitate mechanical operation. The mechanical motion gives an output in terms of linear, rotatory or oscillatory motion. As liquids are nearly impossible to compress, a hydraulic actuator can exert a large force. The drawback of this approach is its limited acceleration.

The hydraulic cylinder consists of a hollow cylindrical tube along which a piston can slide. The term *single acting* is used when the fluid pressure is applied to just one side of the piston. The piston can move in only one direction, a spring being frequently used to give the piston a return stroke. The term *double acting* is used when pressure is applied on each side of the piston; any difference in pressure between the two sides of the piston moves the piston to one side or the other.

Pneumatic rack and pinion actuators for valve controls of water pipes

Pneumatic

Pneumatic actuators enable considerable forces to be produced from relatively small pressure changes. A pneumatic actuator converts energy formed by vacuum or compressed air at high pressure into either linear or rotary motion. Pneumatic energy is desirable for main engine controls because it can quickly respond in starting and stopping as the power source does not need to be stored in reserve for operation. Moreover, pneumatic actuators are safer, cheaper, and often more reliable and powerful than other actuators. These forces are often used with valves to move diaphragms to affect the flow of air through the valve.

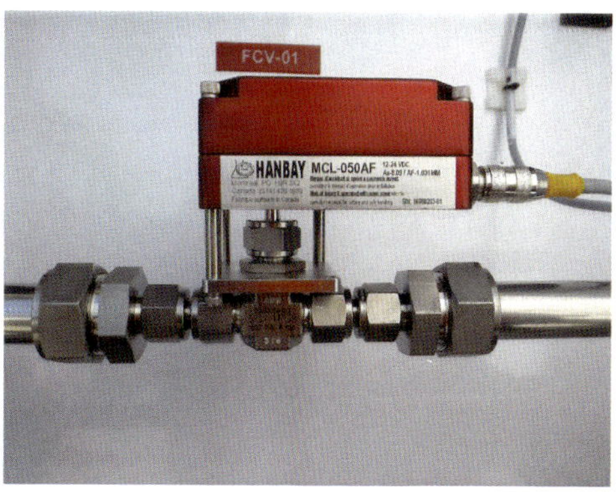

Electric valve actuator controlling a ½ needle valve.

Electric

An electric actuator is powered by a motor that converts electrical energy into mechanical torque. The electrical energy is used to actuate equipment such as multi-turn valves. Additionally, a brake is typically installed above the motor to prevent the media from opening valve. If no brake is installed, the actuator will uncover the opened valve and rotate it back to its closed position. If this continues to happen, the motor and actuator will eventually become damaged. It is one of the cleanest and most readily available forms of actuator because it does not directly involve oil or other fossil fuels.

Twisted and Coiled Polymer (TCP) or Supercoiled Polymer (SCP)

Twisted and coiled polymer (TCP) actuator also known as supercoiled polymer (SCP) actuator is a coiled polymer that can be actuated by electric power. A TCP actuator look like a helical spring. TCP actuators are usually made from silver coated Nylon. TCP actuators can also be made from other electrical conductance coat such as gold. TCP actuator should be under a load to keep the muscle extended. The electrical energy transforms to thermal energy due to electrical resistance, which is also known as Joule heating, Ohmic heating, and resistive heating. As the temperature of the TCP actuator increases by Joule heating, the polymer contracts and it causes the actuator contraction.

Thermal or Magnetic (Shape Memory Alloys)

Actuators which can be actuated by applying thermal or magnetic energy have been used in commercial applications. Thermal actuators tend to be compact, lightweight, economical and with high power density. These actuators use shape memory materials (SMMs), such as shape memory alloys (SMAs) or magnetic shape-memory alloys (MSMAs). Some popular manufacturers of these devices are Finnish Modti Inc., American Dynalloy and Rotork.

Mechanical

A mechanical actuator functions to execute movement by converting one kind of motion, such as rotary motion, into another kind, such as linear motion. An example is a rack and pinion. The operation of mechanical actuators is based on combinations of structural components, such as gears and rails, or pulleys and chains.

3D printed Soft Actuators

Soft actuators are being developed to handle fragile objects like fruit harvesting in agriculture or manipulating the internal organs in biomedicine that has always been a challenging task for robotics. Unlike conventional actuators, soft actuators produce flexible motion due to the integration of microscopic changes at the molecular level into a macroscopic deformation of the actuator materials.

The majority of the existing soft actuators are fabricated using multistep low yield processes such as micro-moulding, solid freeform fabrication, and mask lithography. However, these methods require manual fabrication of devices, post processing/assembly, and lengthy iterations until maturity in the fabrication is achieved. To avoid the tedious and time-consuming aspects of the current fabrication processes, researchers are exploring an appropriate manufacturing approach for effective fabrication of soft actuators. Therefore, special soft systems that can be fabricated in a single step by rapid prototyping methods, such as 3D printing, are utilized to narrow the gap between the design and implementation of soft actuators, making the process faster, less expensive, and simpler. They also enable incorporation of all actuator components into a single structure eliminating the need to use external joints, adhesives, and fasteners. These result in a decrease in the number of discrete parts, post-processing steps, and fabrication time.

3D printed soft actuators are classified into two main groups namely “semi 3D printed soft actuators” and “3D printed soft actuators”. The reason for such classification is to distinguish between the printed soft actuators that are fabricated by means of 3D printing process in whole and the soft actuators whose parts are made by 3D printers and post processed subsequently. This classification helps to clarify the advantages of 3D printed soft actuators over the semi 3D printed soft actuators due to their capability of operating without the need of any further assembly.

Shape memory polymer (SMP) actuators are the most similar to our muscles, providing a response to a range of stimuli such as light, electrical, magnetic, heat, pH, and moisture changes. They have some deficiencies including fatigue and high response time that have been improved through the introduction of smart materials and combination of different materials by means of advanced fabrication technology. The advent of 3D printers has made a new pathway for fabricating low-cost and fast response SMP actuators. The process of receiving external stimuli like heat, moisture, electrical input, light or magnetic field by SMP is referred to as shape memory effect (SME). SMP exhibits some rewarding features such a low density, high strain recovery, biocompatibility, and biodegradability.

Photopolymer/light activated polymers (LAP) are another type of SMP that are activated by light stimuli. The LAP actuators can be controlled remotely with instant response and, without any physical contact, only with the variation of light frequency or intensity.

A need for soft, lightweight and biocompatible soft actuators in soft robotics has influenced researchers for devising pneumatic soft actuators because of their intrinsic compliance nature and ability to produce muscle tension.

Polymers such as dielectric elastomers (DE), ionic polymer metal composites (IPMC), ionic electroactive polymers, polyelectrolyte gels, and gel-metal composites are common materials to form 3D layered structures that can be tailored to work as soft actuators. EAP actuators are categorized as 3D printed soft actuators that respond to electrical excitation as deformation in their shape.

Examples and Applications

In engineering, actuators are frequently used as mechanisms to introduce motion, or to clamp an object so as to prevent motion. In electronic engineering, actuators are a subdivision of transducers. They are devices which transform an input signal (mainly an electrical signal) into some form of motion.

Examples of Actuators

- Comb drive
- Digital micromirror device
- Electric motor
- Electroactive polymer
- Hydraulic cylinder
- Piezoelectric actuator
- Pneumatic actuator
- Screw jack
- Servomechanism
- Solenoid
- Stepper motor
- Shape-memory alloy
- Thermal bimorph
- Hydraulic actuators

Circular to Linear Conversion

Motors are mostly used when circular motions are needed, but can also be used for linear applications by transforming circular to linear motion with a lead screw or similar mechanism. On the other hand, some actuators are intrinsically linear, such as piezoelectric actuators. Conversion between circular and linear motion is commonly made via a few simple types of mechanism including:

- Screw: Screw jack, ball screw and roller screw actuators all operate on the principle of the simple machine known as the screw. By rotating the actuator's nut, the screw shaft moves in a line. By moving the screw shaft, the nut rotates.

- Wheel and axle: Hoist, winch, rack and pinion, chain drive, belt drive, rigid chain and rigid belt actuators operate on the principle of the wheel and axle. By rotating a wheel/axle (e.g. drum, gear, pulley or shaft) a linear member (e.g. cable, rack, chain or belt) moves. By moving the linear member, the wheel/axle rotates.

Virtual Instrumentation

In virtual instrumentation, actuators and sensors are the hardware complements of virtual instruments.

Performance Metrics

Performance metrics for actuators include speed, acceleration, and force (alternatively, angular speed, angular acceleration, and torque), as well as energy efficiency and considerations such as mass, volume, operating conditions, and durability, among others.

Force

When considering force in actuators for applications, two main metrics should be considered. These two are static and dynamic loads. Static load is the force capability of the actuator while not in motion. Conversely, the dynamic load of the actuator is the force capability while in motion.

Speed

Speed should be considered primarily at a no-load pace, since the speed will invariably decrease as the load amount increases. The rate the speed will decrease will directly correlate with the amount of force and the initial speed.

Operating Conditions

Actuators are commonly rated using the standard IP Code rating system. Those that are rated for dangerous environments will have a higher IP rating than those for personal or common industrial use.

Durability

This will be determined by each individual manufacturer, depending on usage and quality.

MEMS Thermal Actuators

Thermal actuation has been extensively employed in MEMS. It includes a broad spectrum of principles such as thermal pneumatic, shape memory alloy (SMA) effect, bimetal effect, mechanical thermal expansion, etc. The thermal pneumatic microactuator uses thermal expansion of a gas or liquid or the phase change between liquid and gas to create the actuation.

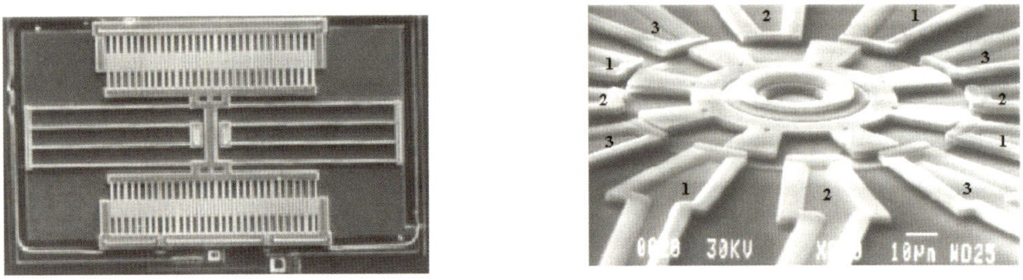

(a) (b)

Figure: (a) Comb-drive electrostatic microactuator. (b) electrostatic micromotor

Example

As shown in Figures, a thermal pneumatic actuator is made of a cavity that contains a volume of fluid with a thin membrane as one wall. Current passed through a heating resistor causes the liquid in the cavity to expand and deform the membrane.

Shape memory alloy effect occurs in some alloys in which a reversible thermal mechanical transformation of the atomic structure of the metal happens at a certain temperature. At low temperature, the SMA is kept at the desired deformed shape. When the temperature rises above a threshold value, the deformed SMA is transformed back to the original shape. A thermal bimetallic microactuator consists of two different materials that are layered together. Figure shows a cantilever bimetallic structure. When it is heated, a deflection is generated by the different thermal expansion between the two materials. The more different the two materials' thermal expansion coefficients, the more deflection is generated. The principle of the mechanical thermal expansion microactuator is similar to that of the bimetallic microactuator. The only difference is that the mechanical thermal expansion microactuators are made of the same material.

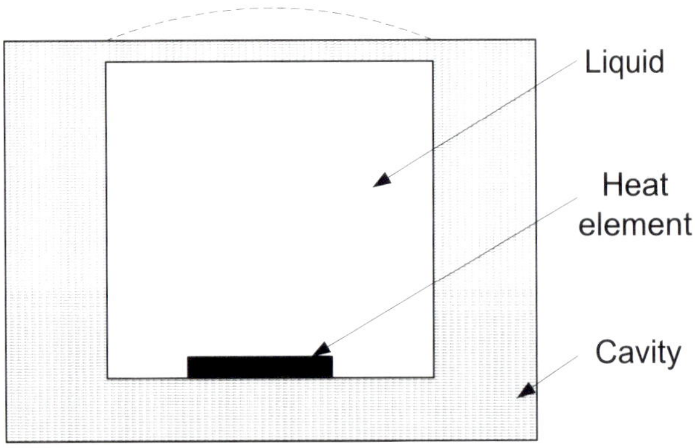

Figure: The schematic diagram of thermal pneumatic microactuator

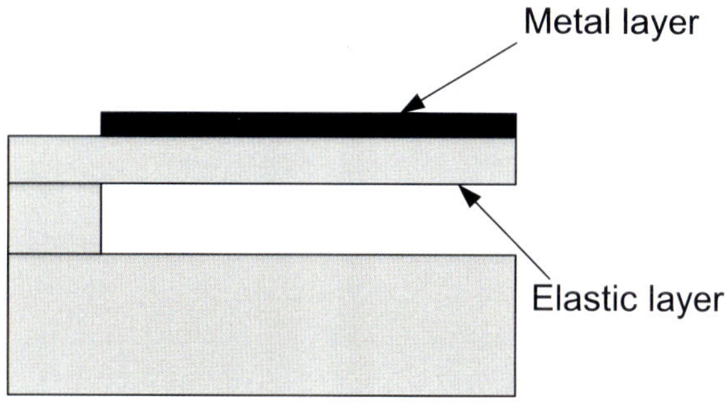

Figure: The thermal bimetallic microactuator with the cantilever prototype

Thermal actuators can generate relatively large force and displacement at low actuating voltage. The deflection can linearly increase as the control voltage is increased within a large range. Mechanical thermal expansion actuator and bimetallic actuator also can be integrated in a chip easily. However the high power consumption and low switching frequency are concerns for applications of thermal actuators.

Micro Magnetic Actuators

An actuator is a machine component, which is responsible for the control and movement of a system or mechanism. A MEMS magnetic actuator uses a microelectromechanical system, which applies the theory of magnetism or Lorentz force equation, to convert an electrical current into some mechanical output.

Operations in MEMS

The three major operations in MEMS are:

- Sensing: Measuring a mechanical input by converting it to an electrical signal, e.g. a MEMS accelerometer or a pressure sensor (could also measure electrical signals as in the case of current sensors)
- Actuation: Using an electrical signal to cause the displacement (or rotation) of a mechanical structure, e.g. a synthetic jet actuator.
- Power generation: Generates power from a mechanical input, e.g. MEMS energy harvesters

These three operations require some form of transduction schemes, the most popular ones being: piezoelectric, electrostatic, piezoresistive, electrodynamic, magnetic and magnetostrictive. The MEMS magnetic actuators use the last three schemes for their operation.

Magnetic Actuation

The principle of magnetic actuation is based on the Lorentz Force Equation.

$$\vec{F}_{mag} = q\vec{v} \times B$$

When a current-carrying conductor is placed in a static magnetic field, the field produced around the conductor interacts with the static field to produce a force. This force can be used to cause the displacement of a mechanical structure.

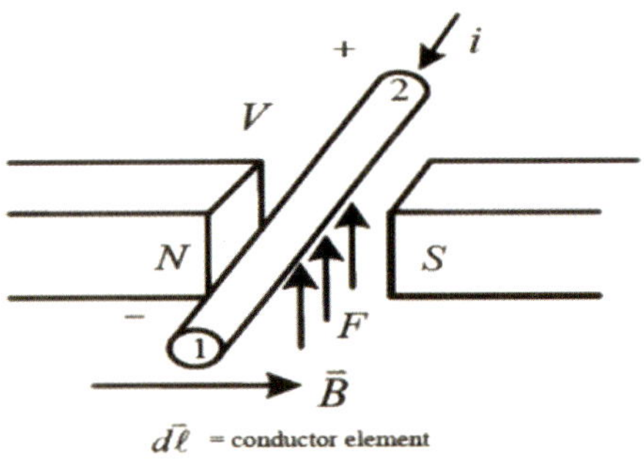

Magnetic actuation

Governing Equations and Parameters

A typical MEMS actuator is shown on the right. For a single turn of circular coil, the equations that govern its operation are:

- The H-field from a circular conductor:

$$H(z) = \frac{Ir^2}{2(r^2 + z^2)^{3/2}}$$

- The force produced by the interaction of the flux densities:

$$F_z = B_l A_{mag} \int_z^{z+h_{mag}} \frac{dHz}{dz} dz$$

The deflection of a mechanical structure for actuation depends on certain parameters of the device. For actuation, there has to be an applied force and a restoring force. The applied force is the force represented by the equation above, while the restoring force is fixed by the spring constant of the moving structure.

The applied force depends on both the field from the coils and the magnet. The remanence value of the magnet, its volume and position from the coils all contribute to its effect on the applied Force. Whereas the number of turns of coil, its size (radius) and the amount of current passing through it determines its effect on the Applied Force.

The spring constant depends on the Young's modulus of the moving structure, and its length, width and thickness.

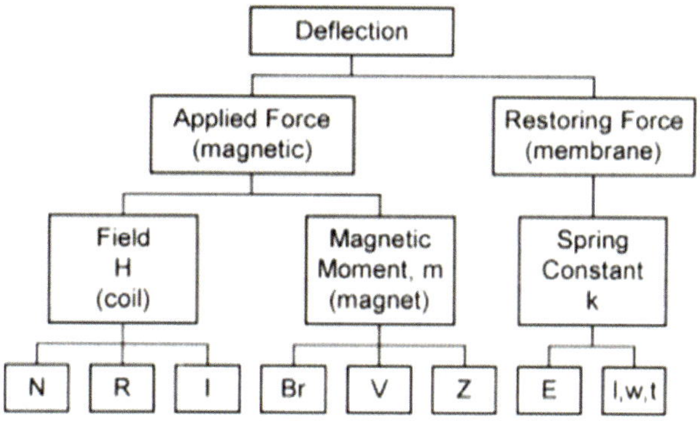

Magnetostrictive Actuators

Magnetic actuation is not limited to the use of Lorentz force to cause a mechanical displacement. Magnetostrictive actuators can also use the theory of magnetism to bring about displacement. Materials that change their shapes when exposed to magnetic fields can now be used to drive high-reliability linear motors and actuators.. An example is a nickel rod that tends to deform when it is placed in an external magnetic field. Another example is wrapping a series of electromagnetic induction coils around a metal tube in which a Terfenol-D material is placed. The coils generate a moving magnetic field that courses wavelike down the successive windings along the stator tube. As the traveling magnetic field causes each succeeding cross section of Terfenol-D to elongate, then contract when the field is removed, the rod will actually "crawl" down the stator tube like an inchworm. Repeated propagating waves of magnetic flux will translate the rod down the tube's length, producing a useful stroke and force output. The amount of motion generated by the material is proportional to the magnetic field provided by the coil system, which is a function of the electric current. This type of motive device, which features a single moving part, is called an elastic-wave or peristaltic linear motor.

Advantages of Magnetic Actuators

- High actuation force and stroke (displacement)
- Direct, fully linear transduction (in the case of electrodynamic actuation)
- Bi-directional actuation
- Contactless remote actuation
- Low-voltage actuation
- A figure of merit for actuators is the density of field energy that can be stored in the gap between the rotor and stator. Magnetic actuation has a potentially high energy density.

Magnet Material

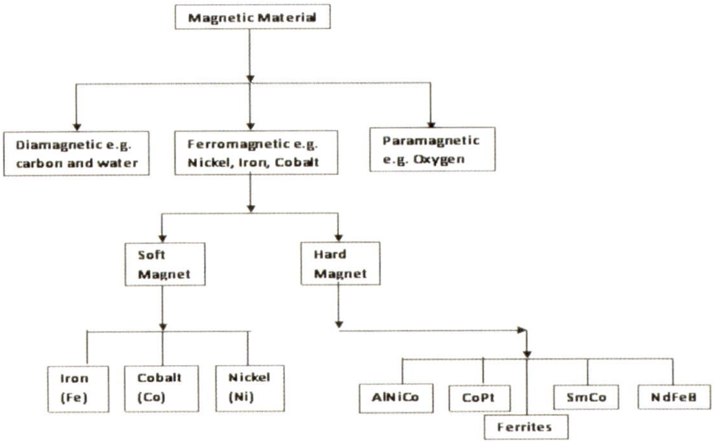

Magnet material selection for static B-Field

The operation of the magnetic actuator depends on the interaction between the field from an electromagnet and a static field. To produce this static field, it is important to use the right material. In MEMS, permanent magnets have become the favorite because they have a very good scaling factor and they retain their magnetization even when there is no external field meaning that they need not be continuously magnetized when they are in use

Integrating the Magnet into the MEMS Device

MEMS devices are designed and fabricated using special micro-fabrication techniques. The major challenge however for magnetic MEMS is the integration of the magnet into the MEMS device. Recent research has suggested solutions to this challenge.

Fabrication (or Molding) of the Magnet

There are several ways by which the magnet could be fabricated on a MEMS structure:

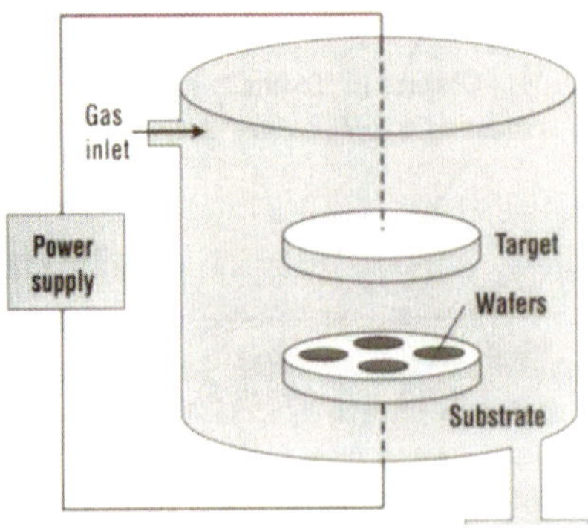

Sputtering

- Sputtering: Argon ion bombardment of the material release particles of the material. Mainly for depositing rare earth magnets. Deposition rate and film surface area depend on sputtering tool and target size

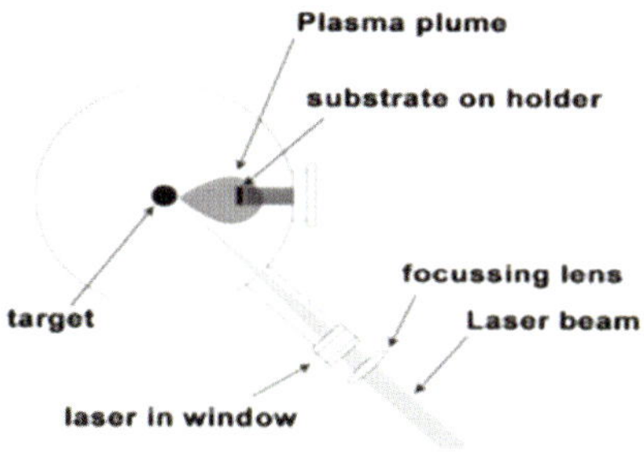

Pulsed Layer Deposition

- Pulsed layer deposition: A high-power pulsed laser beam is focused inside a vacuum chamber to strike a target of the material that is to be deposited
- Electroplating
- Screen printing
- Wax/parylene bonding

Issues with Magnetic Actuation

- High-power dissipation. This is a major problem for magnetic MEMS, but work is underway to circumvent this.
- Fabrication of the coil
- Integration of the micromagnet into the MEMS device
- Process-material compatibility
- Integratability into the overall microfabrication process (maintain cost and throughput)
- So that preexisting processes in the fabrication of the MEMS device will not be tampered with, deposition temperatures and post-deposition treatment/conditions must be tolerable. Also, the micromagnet must be able to withstand any chemical treatment that will come after its deposition
- Issues with magnetization (One may want to have more than one direction of magnetization; this creates a problem)

Each of these challenges can be mitigated or lessened by the right choice of material, choice of molding or fabrication method, and the type of device that is to be constructed. Applications of the magnetic actuator include: the synthetic jet actuator, micro-pumps and micro-relays.

Piezoelectric and Piezoresistive Effect

Piezo comes from the Greek word "piezein," which means "squeeze" or "apply some pressure." Whether they take form as a transducer or sensor, piezo components all operate as the result of some degree of physical pressure placed upon them. Most piezo devices are piezoelectric or piezoresistive, and each has its appropriate applications.

Piezoelectric Effect

Under pressure, vibration, or other forms of stress, piezoelectricity forms in certain materials, particularly crystals. Essentially, the piezoelectric effect is merely the result of stressing a piezo element—crystal, ceramic, or biological matter—to generate a charge or voltage.

The piezoelectric effect is linear. The amount of charge generation is proportional to the amount of stress placed upon the piezo material. Interestingly, this effect is reversible. Applying a charge to the piezo material generates a mechanical response or a pulse. As a result, piezoelectric components find employment in sound production and detection apps, voltage and frequency generation, and a plethora of measurement systems.

- The piezo tweeter, a.k.a., horn, on the left provides a cost alternative for standard magnetic-coil speakers on the right. On its own, the piezo horn responds only to high frequencies, eliminating the need for a frequency-compensating crossover network.

One example of this reverse deployment is the highly cost-effective and efficient piezo tweeter or horn found in inexpensive consumer-audio speaker systems in place of the somewhat more expensive magnetic-coil high-frequency driver with a paper cone. A piezo tweeter uses a piezoelectric crystal, which generates a small voltage when subjected to vibration or pressure.

When functioning as a high-frequency reproduction device (tweeter), however, the crystal is subjected to a voltage—the high-voltage audio signal coming from the receiver or amplifier. The crystal deforms in variation with its input signal and audibly reproduces the input.

The inexpensive piezo crystal responds only to high frequencies, around 4 kHz and above, making this arrangement cost-effective. Therefore, it requires no crossover with numerous passive components to operate as a tweeter in a two- or three-way speaker system. Sometimes, depending on the design, placing an inexpensive electrolytic capacitor in series with the piezo horn provides protection by blocking low frequencies that might blow the horn.

- On the left, the piezoelectric material generates a voltage under pressure or vibration. On the right, the piezoelectric effect is reversible by applying voltage to the piezoceramic for a variety of sound- or pulse-generation purposes.

On a historical note, Pierre and Jacques Curie are credited with the discovery and demonstration of the piezoelectric effect in 1880. Notably, the brothers did not realize the converse piezoelectric effect, which Gabriel Lippmann demonstrated in mathematical form around 1881.

Piezoresistive Effect

The piezoresistive effect also involves pressure or stress. However, changes in resistance across the piezo material are the product, not a charge or voltage. It is a change in electrical resistance of a semiconductor material due to mechanical stress.

Probably the most basic piezoresistive devices are, obviously, piezo resistors. Form factors include integrated resistor networks, potentiometers, and accelerometers. Made from semiconductor materials, piezoresistive devices most commonly are used in pressure measurement.

- When pressure is applied to a piezo resistor, depending on the material, its resistance increases.

In 1856, Lord Kelvin noted the change of resistance in mechanically loaded metal devices. Almost 100 years later, C.S. Smith described the piezoresistive effect in silicon and germanium in 1954.

The most common components that rely on the piezoelectric and piezoresistive effects include, but are not restricted to, transducers and sensors. And, most applications are in detection and measurement.

Transducers

Transducers convert energy from one form to another. Stated earlier, piezoelectric transducers work both ways. They can convert mechanical energy such as pressure and vibration to electrical energy like voltage or current. They also can operate in reverse, converting electrical energy into mechanical energy such as sound or vibration.

Piezoresistive transducers convert mechanical energy into proportionate levels of resistance. They do not convert any form of stimulus to a voltage or current, nor are they reverse active like their piezoelectric cousins, meaning they cannot convert resistance levels to some other form of energy. Both piezoelectric and piezoresistive transducers come in a wide variety of shapes and packages.

Probably the most common piezoelectric component is the disk-shaped variety. The ultra-thin metal disk comes in a range of diameters. Construction is the same for all sizes. Piezo crystals occupy the center portion of the disk. The outer metal circle and body back is the ground for the component.

- The thin metal disk is the most common package for piezoelectric transducers. The piezo crystal or material is housed in the center portion, and the rest of the metal package is the ground.

Hookup is a breeze, requiring just two wires: a hot and a ground. This format also is popular since the thin disk fits into the tightest of quarters. It can be sandwiched between two flat surfaces or simply attached to any flat surface.

Piezoresistive transducers tend to be somewhat larger due to their use of semiconductor materials. However, depending on the application, many housings, sizes, and shapes are available to accommodate compact or larger designs. Since these resistive components only work one way, converting mechanical stimuli into resistance, they find regular employment in pressure-measurement applications.

- Using silicon semiconductor material for the piezoresistive effect, piezoresistive transducers tend to be slighter larger than their piezoelectric counterparts.

Sensors

Sensors detect or measure (or sometimes detect and measure) physical quantities such as distance, pressure, motion, and temperature. They perform their job via conversion. For example, a thermocouple converts a temperature into a readable voltage.

Piezoelectric sensors rely on the piezoelectric effect to measure a plethora of parameters such as pressure, strain, or force by, once again, converting them to voltages. Technically, one can justifiably say that piezoelectric sensors and transducers are one and the same. But piezoelectric sensors, more often than not, operate purely as sensors and not in the aforementioned reverse mode, i.e., applying voltage to generate an effect.

Additionally, piezoelectric sensors are electromechanical components exhibiting near zero deflection. As a result, they respond across a fairly high-frequency bandwidth and exhibit consistent linearity over a wide amplitude range. They are also available in a wide selection of sizes and lengths.

- Maintaining a fairly low profile and available in numerous sizes and lengths, piezoelectric sensors exhibit near zero deflection, making them both rugged and responsive to high frequencies.

Piezoresistive sensors are a bit more sophisticated in their design and the piezo they employ. For instance, they can employ thin metal-film resistors, single-crystal silicon, and other variations. It stands to reason that both the application and the budget will most likely determine which material to choose.

The piezoresistive sensor is a mainstay in pressure-measurement applications. According to Maxim Integrated Products, mono-crystalline silicon pressure sensors have come into wide use lately. Built on semiconductor technology, the resistance change (piezoelectric effect) is notably higher than exhibited in standard strain gauges. Therefore, the sensitivity of mono-crystalline sensors is higher than the sensitivity of most other types.

Available in a wider array of packaging options, piezoresistive sensors offer sensitivities beyond 10 mV/V and stable linearity at constant temperature. They also reliably track pressure changes without hysteresis. Disadvantages include significant nonlinear dependence of the full-scale signal on temperature up to 1%/Kelvin, initial offsets up to 100% of full scale or more, and offset drift with temperature.

Piezoresistive sensors are available in a wider array of packaging options and specify sensitivities greater than 10 mV/V.

References

- Zolfagharian, Ali; Kouzani, Abbas Z.; Khoo, Sui Yang; Gibson, Ian; Kaynak, Akif (2016). "3D printed hydrogel soft actuators". 2016 IEEE Region 10 Conference (TENCON). pp. 2272–7. doi:10.1109/TENCON.2016.7848433. ISBN 978-1-5090-2597-8.
- Electrostatic-discharge-ESD: techtarget.com, Retrieved 16 April 2018
- Zolfagharian, Ali; Kouzani, Abbas Z.; Khoo, Sui Yang; Moghadam, Amir Ali Amiri; Gibson, Ian; Kaynak, Akif (2016). "Evolution of 3D printed soft actuators". Sensors and Actuators A: Physical. 250: 258–72. doi:10.1016/j.sna.2016.09.028.
- "A Great Combination: Pneumatic Actuator, Pneumatic Timer, Pneumatic Valves, and Pneumatic Indicators : Ellis/Kuhnke Controls". www.ekci.com. Archived from the original on 2018-02-21. Retrieved 2018-02-20.
- What-s-difference-between-piezoelectric-and-piezoresistive-components: electronicdesign.com, Retrieved 27 May 2018
- "What is a Pneumatic Actuator?". www.tech-faq.com. Archived from the original on 2018-02-21. Retrieved 2018-02-20.

5

Microelectromechanical Devices

Science and technology have undergone rapid developments in the past decade, which has resulted in innovation in significant tools and techniques in the domain of microelectromechanical devices. This chapter explores topics such as microfluidics, bio-MEMS, digital micromirror devices and MEMS gyroscopes, which will provide an insight into MEMS in the modern scenario.

MEMS Gyroscopes

MEMS gyroscopes are motion sensors that detect and measure the angular motion of an object. It measures the rate of rotation of an object around a particular axis. 1-axis, 2-axis, and 3-axis gyroscopes, capable of measuring velocity in three directions, are available in the market. Currently, 3-axis gyroscopes are combined with 3-axis accelerometers and 3-axis magnetometers to achieve 9 DOF, and these are generally used in consumer electronics. The increased demand for mobile devices has propelled the growth of the MEMS Gyroscopes market. The cost of MEMS gyroscopes is expected to decline drastically during the forecast period, leading to its increased adoption. MEMS Gyroscope is widely used in spacecraft orientation, automotive industry, photography, entertainment, hobbies, industrial robotics, additional applications.

MEM gyroscopes are not a specific type, but a class, of gyroscopes which are printed onto circuit boards using photolithography.

There are many types, but they all rely on the same principle, that of vibrating objects undergoing rotation.

Each type of MEMs gyroscope has some form of oscillating component from where the acccleration, and hence direction change, can be detected. This is because, as per the conservation of motion law, a vibrating object likes to continue vibrating in the same plane, and any vibrational deviation can be used to derive a change in direction.

These deviations are caused by the Corolis force, which is orthogonal to the vibrating object.

MEMS vibrating mass gyroscopes aimed to create smaller and more sensitive devices

Draper Tuning Fork Gyroscope Advancements and Applications

One of the most widely used micro-machined gyroscopes is the tuning fork design from the Charles Stark Draper Lab. The design consists of two tines connected to a junction bar which resonate at certain amplitude.

When the tines rotate, Coriolis force causes a force perpendicular to the tines of the fork. The force is then detected as bending of the tuning fork or a torsional force. These forces are proportional to the applied angular rate, from which the displacements can be measured in a capacitive fashion. Electrostatic, electromagnetic, or piezoelectric mechanisms can be used to detect the force.

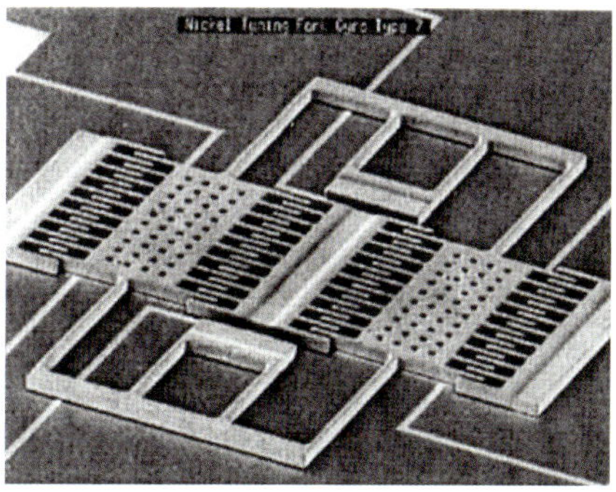

Figure - The first working prototype of the Draper Lab comb drive tuning fork

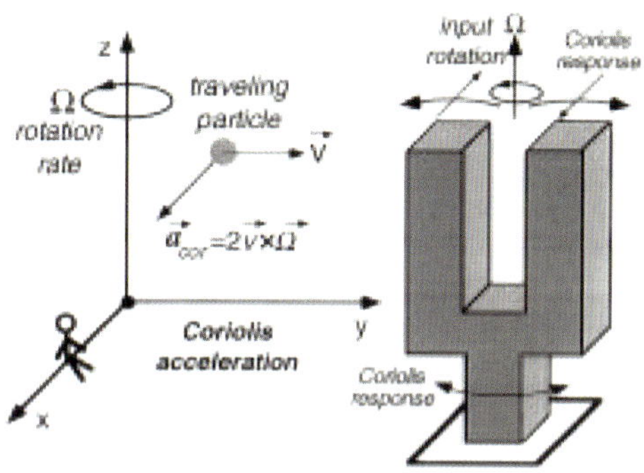

Tuning Fork Physics

Since the development of their first tuning fork gyroscope in 1993, the Draper Laboratory has made significant improvements to the device. Their first gyroscope was developed for the automobile industry. The gyroscope had command of 1 degree/hr drift, and possessed 4000 deg/hr resolution. These devices eventually functioned as the yaw rate sensor for skid control in anti-lock braking applications. Tests run on these sensors involve the examining the change in bias and error of such over a number of variables. Proper data could be retrieved in 0.8 s and sent to the necessary actuator to cause proper breaking in due time. These systems need to operate in a range of temperatures, specifically from -40 to 80 degrees Celsius. Over this range, both the bias error and the scale factor error are both quite stable. The bias error is approximately 2200 deg/h. Scale factor error was approximately 0.08%. Results from these tests are shown in Figures.

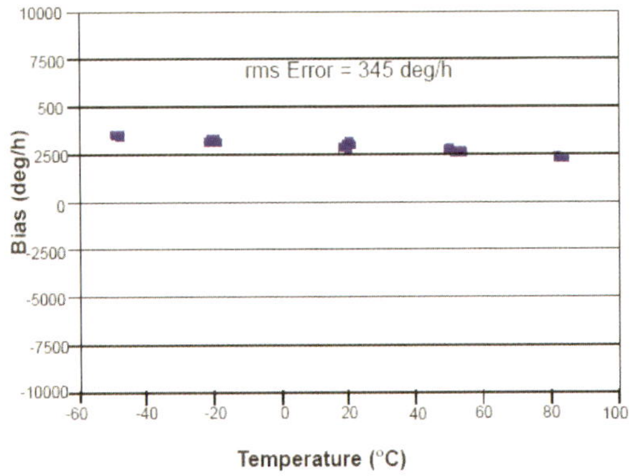

Bias vs. Temperature

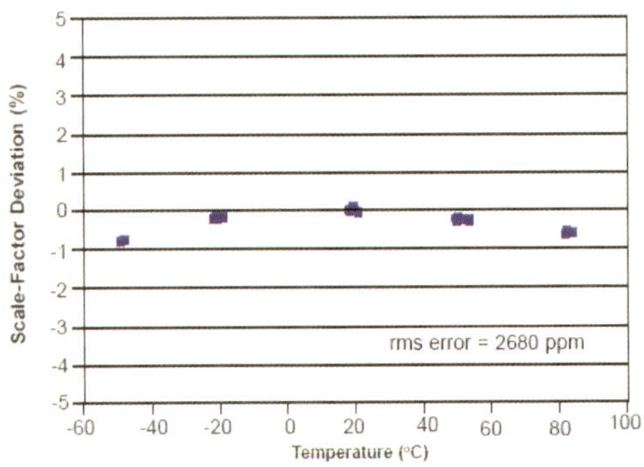

Scale Factor vs.Temperature

Since this initial design, the performance of the tuning fork gyroscope and gradually increased. In 1994, a 500 deg/hr resolution was achieved. The designs in 1997 resulted in resolutions of 100 deg/hr. Drift stability improved an order of magnitude to 0.1 deg/hr. With these types of results, these gyroscopes can be implemented with near exact data replication and production. Along with the increased resolution, the input voltage noise was lowered significantly, leading to a stronger signal-to-noise ratio, providing sensors with the ability to communicate better with their devices.

Piezoelectric Plate Gyroscope

While vibrating ring gyroscopes and tuning fork gyroscopes were the first successful MEMS gyroscopes and are still the most widely produced, other successful MEMS gyroscopes have since been created. One of these gyroscopes is the Piezoelectric Plate Gyroscope which uses a PZT plate as its base. This method, which in the past has been used to try to build macro-scale gyroscopes, is actually ideal for micro devices. At micro levels, an entire plate can be made of piezoelectric material. It has advantages over the common vibrating gyroscopes in that it requires a much smaller drive voltage to create readable outputs.

Physical Description

The piezoelectric plate gyroscope is very simple in its design. In fact, it is much simpler than the ring or fork gyroscopes. There is a piezoelectric plate, which has a length and width much larger than its depth. The plate has electrical leads connected to all 6 sides and sits on top of a thin membrane of a cavity in a silicon wafer. The cavity allows more freedom for the PZT to vibrate and deform. The leads provide the driving voltage and measure the output.

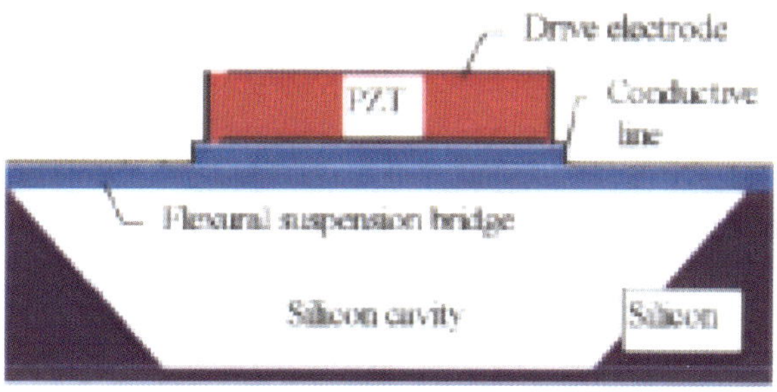

Cross Section of Gyroscope

Fabrication

The following diagrams show a basic fabrication process that could be used to create the piezoelectric gyroscope. To save space not every step is shown.

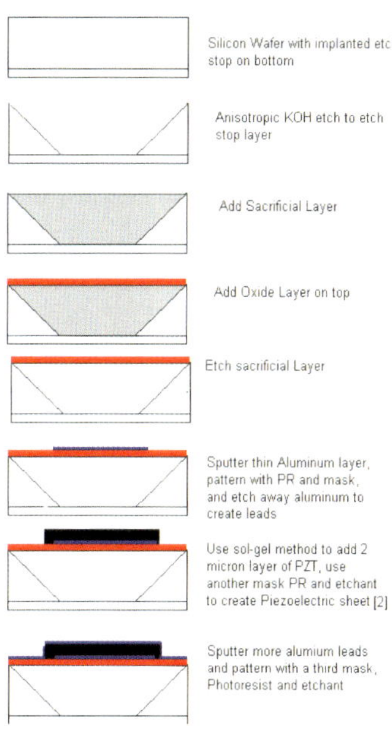

Functional Description

Like other MEMS gyroscope the piezoelectric plate gyroscope works on the principle of a vibrating body. In this case, the vibrating body is a piezoelectric sheet. The sheet does not vibrate like a plate or fork. Instead the thickness vibrates which oscillates with time. This requires an AC driving voltage applied vertically across the plate, which uses the electro-mechanical properties of the PZT to create the vibration. Any piezoelectric material can be used, but PZT has high piezoelectric constants, and can be added at a precise thickness.

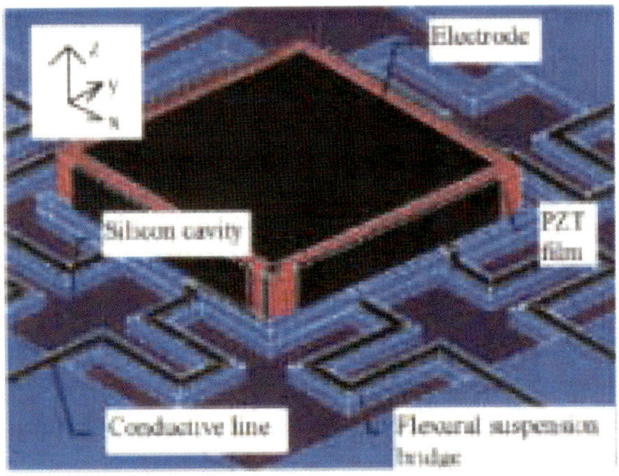

Piezoelectric Plate with orthogonal directions shown

When the vibrating plate is rotated about an axis perpendicular to the drive voltage, a voltage is produced in the third perpendicular direction. This output voltage is proportional to the angular velocity. It can be proven that the relationship is:

$$V_{Out} = -\frac{K}{\epsilon_{11}^{T}}\frac{ac}{4}\left(d_{33} + \left|d_{31}\right|\frac{a^2}{c^2}\right)d_{15}$$

$$\cong \frac{V_{A\rho\,\omega\cos\omega t}}{\epsilon_{11}^{T}}\frac{a^3}{4_c}\left|d_{31}\right|d_{15}\Omega.$$

Here VA is the input voltage magnitude,

ρ is the PZT density, ω is the input voltage frequency, t is time, a is the plate length perpendicular to the direction of rotation, c is the plate thickness, d31 and d15 are piezoelectric constants, ε11 is the permittivity constant and Ω is the angular velocity.

Since the plate has x-y symmetry as shown above, it follows that a single plate can measure rotation in two directions. This is an added advantage when compared to traditional gyroscopes, which only measure one direction of rotation.

Digital Micromirror Device

DLP chip, or Digital Micromirror Device (DMD), is an array of micromirrors that can be used for high speed, efficient, and reliable spatial light modulation. At the heart of every DLP chipset is an array of highly reflective aluminum micromirrors known as the digital micromirror device (DMD). The DMD is an electrical input, optical output micro-electrical- mechanical system (MEMS) that allows developers to perform high speed, efficient, and reliable spatial light modulation. Using TI's proven semiconductor manufacturing capabilities, each DMD contains up to 2 million individually controlled micromirrors built on top of an associated CMOS memory cell.

The DMD project began as the Deformable Mirror Device in 1977 using micromechanical analog light modulators. The first analog DMD product was the TI DMD2000 airline ticket printer that used a DMD instead of a laser scanner.

A DMD chip has on its surface several hundred thousand microscopic mirrors arranged in a rectangular array which correspond to the pixels in the image to be displayed. The mirrors can be individually rotated ±10-12°, to an on or off state. In the on state, light from the projector bulb is reflected into the lens making the pixel appear bright on the screen. In the off state, the light is directed elsewhere (usually onto a heatsink), making the pixel appear dark.

To produce greyscales, the mirror is toggled on and off very quickly, and the ratio of on time to off time determines the shade produced (binary pulse-width modulation). Contemporary DMD chips can produce up to 1024 shades of gray (10 bits).

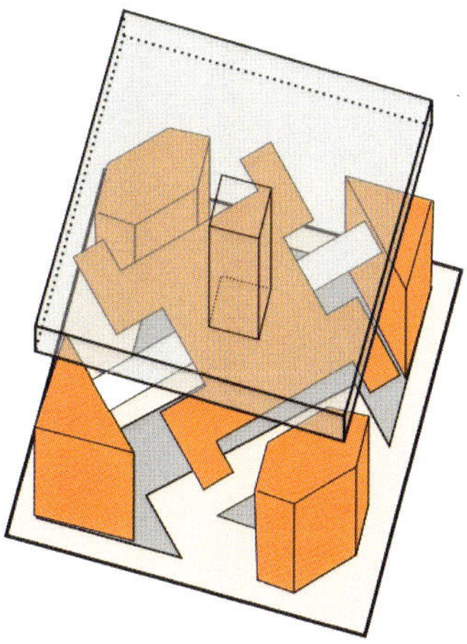

Diagram of a Digital micromirror showing the mirror mounted on the suspended yoke with the torsion spring running bottom left to top right (light grey), with the electrostatic pads of the memory cells below (top left and bottom right)

The mirrors themselves are made out of aluminum and are around 16 micrometers across. Each one is mounted on a yoke which in turn is connected to two support posts by compliant torsion hinges. In this type of hinge, the axle is fixed at both ends and twists in the middle. Because of the small scale, hinge fatigue is not a problem and tests have shown that even 1 trillion (10) operations do not cause noticeable damage. Tests have also shown that the hinges cannot be damaged by normal shock and vibration, since it is absorbed by the DMD superstructure.

Two pairs of electrodes control the position of the mirror by electrostatic attraction. Each pair has one electrode on each side of the hinge, with one of the pairs positioned to act on the yoke and the other acting directly on the mirror. The majority of the time, equal bias charges are applied to both sides simultaneously. Instead of flipping to a central position as one might expect, this actually holds the mirror in its current position. This is because attraction force on the side the mirror is already tilted towards is greater, since that side is closer to the electrodes.

To move the mirrors, the required state is first loaded into an SRAM cell located beneath each pixel, which is also connected to the electrodes. Once all the SRAM cells have been loaded, the bias voltage is removed, allowing the charges from the SRAM cell to prevail, moving the mirror. When the bias is restored, the mirror is once again held in position, and the next required movement can be loaded into the memory cell.

The bias system is used because it reduces the voltage levels required to address the pixels such that they can be driven directly from the SRAM cell, and also because the bias voltage can be removed at the same time for the whole chip, so every mirror moves at the same instant. The advantages of the latter are more accurate timing and a more cinematic moving image.

A broken DMD chip showing the “white dots” appearing on screen as “white pixels”.

Applications

- Televisions and HDTVs.
- Holographic Versatile Discs.
- Head-mounted displays.

- Digital cinema
- DLP projector : Digital Light Processing (DLP).
- Metrology : Optical metrology
- Laser beam machining
- Multivariate Optical Computing.

Scope of Products and Solutions

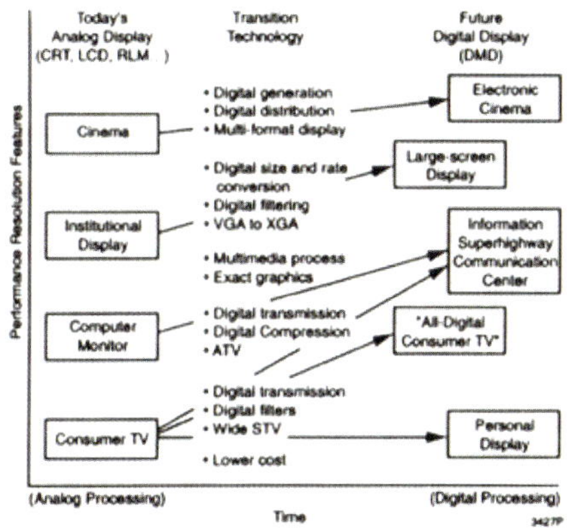

Potential DMD video display systems in today's markets range from conventional consumer TVs to institutional high-end projectors and electronic cinema, as shown in figure.

Any application domain is possible; however, the inherent nature of a digital display tends toward high-performance applications, such as highdefinition or high-quality displays. In addition, the digital nature of the technology matches well with today's surge in computer graphics display. In contrast, rapid integration of digital semiconductor technology could enable low-cost applications of the technology, such as consumer TV or personal viewers. While digital displays have inherent high performance, this can be traded for a cost advantage in these applications. Future displays will focus on the display of media from several sources, such as digital multimedia TVs or personal ports into the "information superhighway." In general, the scope of the DMD display application space is very large and seemingly ever- increasing high- definition or high-quality displays. In addition, the digitalnature of the technology matches well with today's surge in computer graphics display. In contrast, rapid integration of digital semiconductor technology could enable low-cost applications of the technology, such as consumer TV or personal viewers. While digital displays have inherent high performance, this can be traded for a cost advantage in these applications. Future displays will focus on the display of media from several sources, such as digital multimedia TVs or personal ports into the "information superhighway." In general, the scope of the DMD display application space is very large and seemingly ever- increasing.

Digital Processing Impact

The advantages of DMD technology for projection display as:

- Exceptional visual quality display, free from limitations of other technologies:
    - High light throughput efficiency for very bright projection display.
    - Inherent convergence of single DMD systems.
    - Pure color fidelity with pixelated micromechanical mirrors.
    - Electromagnetic radiation-free.
    - High contrast ratio (with use of hidden hinge technology.
    - No persistence, comet tails, or lag.
    - No warmup time.
- Inherently clean digitally generated graphics display.
- Flexible design for product differentiation:
    - Compact form (12 to 14 inch depths possible). o Rapid front and rear projection mode change. o Simpler optics solutions.
    - Less heat sensitivity for high-brightness applications (versus LCD).
- Semiconductor technologies:
    - Economies of scale enable new markets such as large-screen HDTV.
    - Reliability.

While these represent many compelling advantages of DMD and suggest true advantages for all video display applications, regardless of the nature of the video signals, the fact that we can now construct a *fully* digital video display system, with both digital video processing and display, has additional advantages. With the advent of this *truly* digital display, it seems intuitive to use digital video processing to create, transmit, and manipulate the video signals. Although it is difficult for a new display technology to change the infrastructure of an analog industry many decades old, several factors point to a convergence on digital processing in the video industry in the near future:

- Acceptance of the recently defined *all-digital* advanced television (ATV) standard for high-definition TV broadcast.
- Merging of computer graphics with television (i.e., Modular Windows on TVs and TV pictures in Windows).
- The eventual coupling of digital cable and telephone networks.

- Evolution of billion-operation-per-second microprocessors and high-density memories.
- Increasing use of digital processing technology in TV and film production.

Only DMD technology permits digital processing *and* digital display of video. Images can be created, processed, distributed, stored, and displayed all in completely digital form. With DMD displays, the viewer sees extremely sharp and crisp digital images, a result of using time integration of bit images at rates exceeding human perception to create shades of gray (or color). To the system, the DMD appears as a digital memory. Redefining the architecture of the entire video system decreases costs and increases performance from current analog display technologies (CRT and LCD).

Advantages of Completely Digital Video Display

Using an integrated digital display in a projection system, rather than an analog display, maximizes the impact of the system's digital processing, yielding clean video images and reduced system cost. The advantages of implementing a completely digital video projection display system are:

- Inherent noise immunity.
- Stable, consistent operation, independent of temperature, signal, or component lifetime variations.
- Simple product differentiation (with performance and cost variations) by the selection of digital functions.
- Enhanced features and video quality with use of computer video algorithms.
- Unique new features only possible with digital video processing in the display.

With all-digital television sets or computer displays, users can watch without annoying ground- loop no ise or electromagnetic interference from household appliances or local radiation sources. By maintaining digital processing throughout the video system, an architecture can be composed to add functionality without signal degradation. In some cases, the functionality can be altered by simple parametric or software changes. For example, by using programmable processors such as the scanline video processor (SVP) or multimedia video processor (MVP) from TI, we can alter the function of the video processor by simply changing the software loaded via disk, memory card, or networks (cable or telephone). Most algorithms (such as de- interlace, scaling, and sharpness) have functionality that is related to the complexity of the processing and, therefore, to the cost of performing those functions. This ability to optimize the algorithms to the product, including the type and mix of functions, is a feature of digital processing. Changes in analog systems are generally more complex and lead to component or board selection or removal. In addition, digital architectures can be made modular, enabling ease of "drop- in" additional functionality, leading to upgradable systems. This modular approach could be particularly attractive in today's fast changing video environment, since the customer would not need to commit to a single system.

Enhanced Features

Computer video algorithms that permit enhanced features and video quality in a digital system include the following:

- Nonlinear processing to remove distribution artifacts.
- Adaptive filters to improve perceived display resolution.
- Higher performance adaptive digital sharpness control (current sharpness controls are only horizontal).
- Dynamic range enhancement algorithms.
- Dithering to reduce number of bits/pixel transmitted or processed.
- Scene-adaptive histogram specification.
- De-interlace algorithms such as motion-adaptive or edge- interpolative.
- Simple de- interlace algorithms such as line-double or median filter.
- Digital scaling for resizing and aspect ratio conversion: bi- linear to custom-designed FIR filters.
- Vertical aperture correction.
- Features such as PIP, POP, freeze frame, zoom.
- Reduction of display artifacts such as cross-color, dot interference.
- Frame rate conversion.

In addition, the digital video display system uniquely opens new applications and features. These include:

- Interactive communication and advanced video games.
- Sophisticated computer- like interactive user interfaces.
- User video editing (program selection and commercial cuts).
- On-line user-customized video and audio.
- New uses such as video phones, Photo CDs™, and video printers. These features highlight the wide variation of performance and cost possible with digital video processing.

Direct Coupling DMD to Digital Video Systems

Eliminating the analog-to-digital and/or digital-to-analog conversion process for all digital systems and/or relocating analog functions to the digital domain for mixed analog/digital systems lowers the cost and increases the performance of DMD systems. In digital compressed video input systems, the added cost and noise injected by convening to analog signals is not needed for the digital display. Directly digital coupling of the display system to the digital transmission or digital storage media, such as when implementing MPEG-based cable decoders or MPEG-2- based ATV decoders, will remove a source of noise and cost. In particular, more bits per pixel or dynamic range will be available for less cost to the consumer since expensive high-precision ADCs will be in the studio rather than in the home receiver. In addition, direct decoding these digital bitstreams to formatted DMD video, rather than NTSC video, will improve those systems. In "mixed" analog and digital systems, moving functions such as sharpness and picture controls to the digital domain will yield more flexible and repeatable control of those operations.

Impact of DMD on the Display System Electronics

Use of the DMD greatly affects the entire video system electronics. Just as the characteristics of CRTs have influenced TV electronics, we must adapt the system electronics to realize the full benefits of the new features of the DMD display. Use of the DMD can require unique digital signal processing to prepare the data for display of video, including:

- Digital Video Processing — Minimize distribution, storage, and display artifacts.
- DMD Data Formatting — Spatial and temporal change of video for DMDs.
- De-Gamma and Artifact Concealment — Picture linearizing manipulation and minimization of contours.
- Lower Frame Rate Display — Without large-area flicker.
- Synchronization to Nonstandard Video — Video sources vary widely in practice.

DMD Data Formatting

With DMDs, we use a common digital processing function, pulsewidth modulation (PWM), to create a unique display method. In particular, the pixel (or byte) sequential data format used by most display systems must be convened to a bit-plane parallel data format for display on current DMDs. By use of electronic control signals and an integral memory array, the DMD's array of mirrors is time modulated with bit images or patterns that correspond to a desired "accumulated" image. Basically, the eye integrates a sequence of binary images with differing display times (although within the frame time). The display times of each binary image are weighted by the binary code (i.e., in an 8 bit system, 128 would require an "on" pixel for 8.3 ms and an "off" pixel for 8.3 ms of the 16.6 ms frame). Understanding this data formatting, the mechanism for concealing display artifacts, and the overall impact of processing for a sharp, linear display can lead to optimal systems for digital display. In effect, we have exploited the properties and limitations of the human visual system to develop optimized data formatting.

Microfluidics

Microfluidics is the study of systems that can process small quantities of fluids by using tiny channels having dimensions at the microscale - typically tens to hundreds of micrometres. Although in the nascent stage, microfluidics is rapidly emerging as a breakthrough technology that finds applications in diverse fields ranging from biology and chemistry to information technology and optics.

As more and more researchers realize the power of microfluidics, the technology is being applied to many new fields thus saving money and time spent in research. Despite the compelling advantages of microfluidics, it is not being widely used yet - this is possibly because of the challenges in the commercialization of this technology.

Microfluidics and Photolithography

Microfluidic systems are developed using a simple technique called photolithography, which was originally developed for creating small features on circuits in the semiconductor industry. Photolithography is the process used to transfer geometric shapes present on a mask to the surface of a suitable substrate. It makes use of special polymers that react to specific wavelengths of light to create the desired geometric patterns on a substrate.

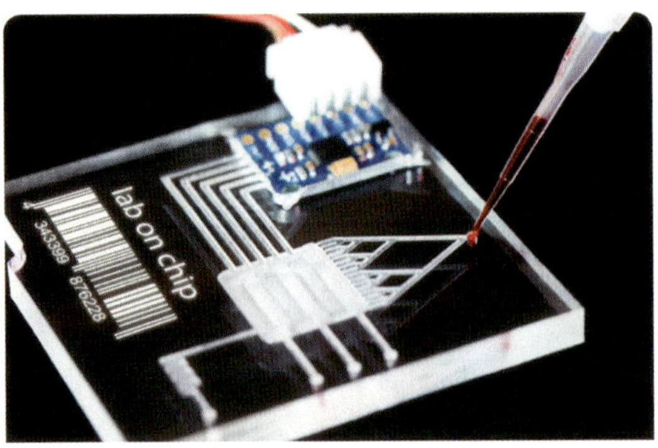

Polymers such as polydimethylsiloxane (PDMS) have replaced silicon and glass and are widely used in the photolithographic production of microfluidic devices. PDMS is a transparent elastic polymer permeable to oxygen and carbon dioxide and hence is useful for housing cells. Once a mold is created, several tiny devices can be produced using the mold and they can be applied in research and diagnostics.

Microscale Behaviour of Fluids

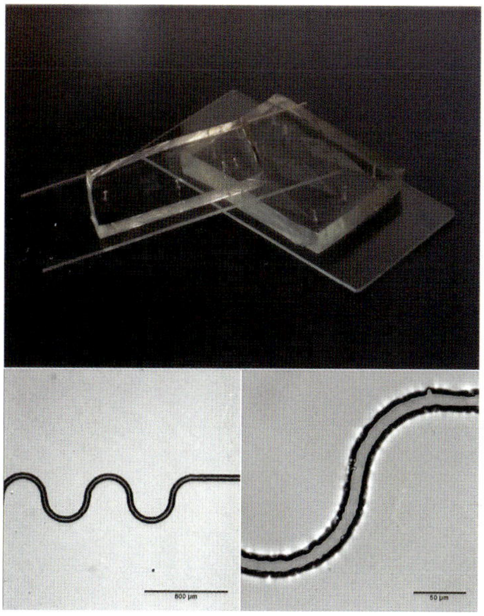

Silicone rubber and glass microfluidic devices. Top: a photograph of the devices. Bottom: Phase contrast micrographs of a serpentine channel ~15 μm wide.

The behaviour of fluids at the microscale can differ from "macrofluidic" behaviour in that factors such as surface tension, energy dissipation, and fluidic resistance start to dominate the system. Microfluidics studies how these behaviours change, and how they can be worked around, or exploited for new uses.

At small scales (channel size of around 100 nanometers to 500 micrometers) some interesting and sometimes unintuitive properties appear. In particular, the Reynolds number (which compares the effect of the momentum of a fluid to the effect of viscosity) can become very low. A key consequence is co-flowing fluids do not necessarily mix in the traditional sense, as flow becomes laminar rather than turbulent; molecular transport between them must often be through diffusion.

High specificity of chemical and physical properties (concentration, pH, temperature, shear force, etc.) can also be ensured resulting in more uniform reaction conditions and higher grade products in single and multi-step reactions.

Key Application Areas

Microfluidic structures include micropneumatic systems, i.e. microsystems for the handling of off-chip fluids (liquid pumps, gas valves, etc.), and microfluidic structures for the on-chip handling of nanoliter (nl) and picoliter (pl) volumes. To date, the most successful commercial application of microfluidics is the inkjet printhead. Additionally, advances in microfluidic manufacturing allow the devices to be produced in low-cost plastics and part quality may be verified automatically.

Advances in microfluidics technology are revolutionizing molecular biology procedures for enzymatic analysis (e.g., glucose and lactate assays), DNA analysis (e.g., polymerase chain reaction and high-throughput sequencing), and proteomics. The basic idea of microfluidic biochips is to integrate assay operations such as detection, as well as sample pre-treatment and sample preparation on one chip.

An emerging application area for biochips is clinical pathology, especially the immediate point-of-care diagnosis of diseases. In addition, microfluidics-based devices, capable of continuous sampling and real-time testing of air/water samples for biochemical toxins and other dangerous pathogens, can serve as an always-on "bio-smoke alarm" for early warning.

Microfluidic technology has led to the creation of powerful tools for biologists to control the complete cellular environment, leading to new questions and discoveries. Many diverse advantages of this technology for microbiology are listed below:

- General single cell studies including growth
- Cellular aging: microfluidic devices such as the "mother machine" allow tracking of thousands of individual cells for many generations until they die.
- Microenvironmental control: Ranging from mechanical environment to chemical environment
- Precise spatiotemporal concentration gradients by incorporating multiple chemical inputs to a single device
- Force measurements of adherent cells or confined chromosomes: Objects trapped in a microfluidic device can be directly manipulated using optical tweezers or other force-generating methods
- Confining cells and exerting controlled forces by coupling with external force-generation methods such as Stokes flow, optical tweezer, or controlled deformation of the PDMS device
- Electric field integration
- Plant on a chip and plant tissue culture
- Antibiotic resistance: Microfluidic devices can be used as heterogeneous environments for microorganisms. In a heterogeneous environment, it is easier for a microorganism to evolve. This can be useful for testing the acceleration of evolution of a microorganism / for testing the development of antibiotic resistance.

Open Microfluidics

In open microfluidics, at least one boundary of the system is removed, exposing the fluid to air or another interface (i.e. liquid). Advantages of open microfluidics include accessibility to the flowing liquid for intervention, larger liquid-gas surface area, and minimized bubble formation. Another advantage of open microfluidics is the ability to integrate open systems with surface-tension driven fluid flow, which eliminates the need for external pumping methods such as peristaltic or syringe pumps. Open microfluidic devices are also easy and inexpensive to fabricate by milling, thermoforming, and hot embossing. In addition, open microfluidics eliminates the need to glue or bond a cover for devices which could be detrimental for capillary flows. Examples of open microfluidics include open-channel microfluidics, rail-based microfluidics, paper-based, and thread-based microfluidics. Disadvantages to open systems include susceptibility to evaporation, contamination, and limited flow rate.

Continuous-flow Microfluidics

These technologies are based on the manipulation of continuous liquid flow through microfabricated channels. Actuation of liquid flow is implemented either by external pressure sources, external mechanical pumps, integrated mechanical micropumps, or by combinations of capillary forces and electrokinetic mechanisms. Continuous-flow microfluidic operation is the mainstream approach because it is easy to implement and less sensitive to protein fouling problems. Continuous-flow devices are adequate for many well-defined and simple biochemical applications, and for certain tasks such as chemical separation, but they are less suitable for tasks requiring a high degree of flexibility or fluid manipulations. These closed-channel systems are inherently difficult to integrate and scale because the parameters that govern flow field vary along the flow path making the fluid flow at any one location dependent on the properties of the entire system. Permanently etched microstructures also lead to limited reconfigurability and poor fault tolerance capability. Computer-aided design automation approaches for continuous-flow microfluidics have been proposed in recent years to alleviate the design effort and to solve the scalability problems.

Process monitoring capabilities in continuous-flow systems can be achieved with highly sensitive microfluidic flow sensors based on MEMS technology which offers resolutions down to the nanoliter range.

Droplet-based Microfluidics

Droplet-based microfluidics is a subcategory of microfluidics in contrast with continuous microfluidics; droplet-based microfluidics manipulates discrete volumes of fluids in immiscible phases with low Reynolds number and laminar flow regimes. Interest in droplet-based microfluidics systems has been growing substantially in past decades. Microdroplets allow for handling miniature volumes (μl to fl) of fluids conveniently, provide better mixing, encapsulation, sorting, and sensing, and suit high throughput experiments. Exploiting the benefits of droplet-based microfluidics efficiently requires a deep understanding of droplet generation to perform various logical operations such as droplet motion, droplet sorting, droplet merging, and droplet breakup.

Digital Microfluidics

Alternatives to the above closed-channel continuous-flow systems include novel open structures, where discrete, independently controllable droplets are manipulated on a substrate using electrowetting. Following the analogy of digital microelectronics, this approach is referred to as digital microfluidics. Le Pesant et al. pioneered the use of electrocapillary forces to move droplets on a digital track. The "fluid transistor" pioneered by Cytonix also played a role. The technology was subsequently commercialised by Duke University. By using discrete unit-volume droplets, a microfluidic function can be reduced to a set of repeated basic operations, i.e., moving one unit of fluid over one unit of distance. This "digitisation" method facilitates the use of a hierarchical and cell-based approach for microfluidic biochip design. Therefore, digital microfluidics offers a flexible and scalable system architecture as well as high fault-tolerance capability. Moreover, because each droplet can be controlled independently, these systems also have dynamic reconfigurability, whereby groups of unit cells in a microfluidic array can be reconfigured to change their functionality during the concurrent execution of a set of bioassays. Although droplets are manipulated in confined microfluidic channels, since the control on droplets is not independent, it should not be confused as "digital microfluidics". One common actuation method for digital microfluidics is electrowetting-on-dielectric (EWOD). Many lab-on-a-chip applications have been demonstrated within the digital microfluidics paradigm using electrowetting. However, recently other techniques for droplet manipulation have also been demonstrated using surface acoustic waves, optoelectrowetting, mechanical actuation, etc.

Paper-based Microfluidics

Paper-based microfluidic devices fill a growing niche for portable, cheap, and user-friendly medical diagnostic systems. Paper based microfluidics rely on the phenomenon of capillary penetration in porous media. In order to tune fluid penetration in porous substrates such as paper, in two and three dimensions, the pore structure, wettability and geometry of the microfluidic devices can be controlled while the viscosity and evaporation rate of the liquid play a further significant role. Many such devices feature hydrophobic barriers on hydrophilic paper that passively transport aqueous solutions to outlets where biological reactions take place. Current applications include portable glucose detection and environmental testing, with hopes of reaching areas that lack advanced medical diagnostic tools.

DNA Chips (Microarrays)

Early biochips were based on the idea of a DNA microarray, e.g., the GeneChip DNAarray from Affymetrix, which is a piece of glass, plastic or silicon substrate, on which pieces of DNA (probes) are affixed in a microscopic array. Similar to a DNA microarray, a protein array is a miniature array where a multitude of different capture agents, most frequently monoclonal antibodies, are deposited on a chip surface; they are used to determine the presence and/or amount of proteins in biological samples, e.g., blood. A drawback of DNA and protein arrays is that they are neither reconfigurable nor scalable after manufacture. Digital microfluidics has been described as a means for carrying out Digital PCR.

Molecular Biology

In addition to microarrays, biochips have been designed for two-dimensional electrophoresis, transcriptome analysis, and PCR amplification. Other applications include various electrophoresis and liquid chromatography applications for proteins and DNA, cell separation, in particular, blood cell separation, protein analysis, cell manipulation and analysis including cell viability analysis and microorganism capturing.

Evolutionary Biology

By combining microfluidics with landscape ecology and nanofluidics, a nano/micro fabricated fluidic landscape can be constructed by building local patches of bacterial habitat and connecting them by dispersal corridors. The resulting landscapes can be used as physical implementations of an adaptive landscape, by generating a spatial mosaic of patches of opportunity distributed in space and time. The patchy nature of these fluidic landscapes allows for the study of adapting bacterial cells in a metapopulation system. The evolutionary ecology of these bacterial systems in these synthetic ecosystems allows for using biophysics to address questions in evolutionary biology.

Cell Behavior

The ability to create precise and carefully controlled chemoattractant gradients makes microfluidics the ideal tool to study motility, chemotaxis and the ability to evolve / develop resistance to antibiotics in small populations of microorganisms and in a short period of time. These microorganisms including bacteria and the broad range of organisms that form the marine microbial loop, responsible for regulating much of the oceans' biogeochemistry.

Microfluidics has also greatly aided the study of durotaxis by facilitating the creation of durotactic (stiffness) gradients.

Cellular Biophysics

By rectifying the motion of individual swimming bacteria, microfluidic structures can be used to extract mechanical motion from a population of motile bacterial cells. This way, bacteria-powered rotors can be built.

Optics

The merger of microfluidics and optics is typical known as optofluidics. Examples of optofluidic devices are tunable microlens arrays and optofluidic microscopes.

Microfluidic flow enables fast sample throughput, automated imaging of large sample populations, as well as 3D capabilities. or superresolution.

Acoustic Droplet Ejection (ADE)

Acoustic droplet ejection uses a pulse of ultrasound to move low volumes of fluids (typically nanoliters or picoliters) without any physical contact. This technology focuses acoustic energy into a fluid sample in order to eject droplets as small as a millionth of a millionth of a litre (picoliter = 10^{-} litre). ADE technology is a very gentle process, and it can be used to transfer proteins, high molecular weight DNA and live cells without damage or loss of viability. This feature makes the technology suitable for a wide variety of applications including proteomics and cell-based assays.

Fuel Cells

Microfluidic fuel cells can use laminar flow to separate the fuel and its oxidant to control the interaction of the two fluids without a physical barrier as would be required in conventional fuel cells.

Astrobiology

In order to understand the prospects for life to exist elsewhere in the universe, astrobiologists are interested in measuring the chemical composition of extraplanetary bodies. Because of their small size and wide-ranging functionality, microfluidic devices are uniquely suited for these remote sample analyses. From an extraterrestrial sample, the organic content can be assessed using microchip capillary electrophoresis and selective fluorescent dyes. These devices are capable of detecting amino acids, peptides, fatty acids, and simple aldehydes, ketones, and thiols. These analyses coupled together could allow powerful detection of the key components of life, and hopefully inform our search for functioning extraterrestrial life.

Future Directions

- On-chip characterization.

- Microfluidics in the classroom: On-chip acid-base titrations.

Bio-MEMS

Bio MEMS is a subset of microelectromechanical systems (MEMS) and microtechnology. Bio MEMS applies to biological systems, in general and to human health, in particular. It has emerged as a subset of MEMS devices for applications in biomedical research and medical microdevices.

The techniques and processes from Bio-MEMS can provide a means to control a wide variety of microenvironmental cues in order to better understand the nature of the cellular response. Findings that have come from the use of Bio-MEMS have influenced the design of biomaterials and tissue-engineered constructs because engineers and scientists can begin to incorporate the appropriate interactions for cells.

Cells in living tissue have different levels of cues than their counterparts in culture. The vascular and lymphatic systems provide cells with the appropriate nutrients and hormones for their survival and function. Likewise, the levels of oxygen and growth factors in culture can be excessive for cells and overdrive their response. These chemical interactions influence cell behavior by engaging surface receptors that activate specific signal pathways. Gradients in signal molecules can also serve to attract cells to locations or activate particular responses. In addition to these soluble factors, the insoluble cues that a cell encounters from its extracellular matrix (ECM) are different than those in tissue. Tissue culture dishes are coated usually with one kind of ECM protein such as collagen or fibronectin or reconstituted extracts such as Matrigel, whose contents are poorly defined. These insoluble factors are not well-matched to the different variety of adhesive ligands that a cell type may require. Moreover, cells experience a variety of extracellular forces from their surroundings that act on the mechanosensory structures in a cell that influence its behavior. These forces come from the range of physiological phenomena such as pressure forces from muscle contraction, shear forces from vascular hemodynamics, traction forces from surrounding cells, or stretching forces from musculoskeletal locomotion. Emerging evidence shows that biomechanical factors such as substrate stiffness and cell geometry can influence cell functions such as migration, cytokinesis, differentiation, and growth. These findings are important because they implicate that cells in one kind of environment respond differently than those in another. Since the influence of different factors must be careful examined, robust means to explore these factors are required.

BioMEMS has provided important insights in improving the development of biomaterials or tissue-constructs to better represent the cellular microenvironment, either by closely matching the native one or specifically activating a cellular response. Many of these tools are derived from microelectromechanical systems (MEMS) such as actuators or posts but have been adapted by researchers interested in probing cells at the nanoscale. BioMEMS devices can be used to measure traction forces that cells produce against a surface, which is important to maintain stable adhesion against a substrate. These tools can also be incorporated with techniques to impart forces onto the surface of a cell to examine the mechanotransduction response. Cell culture and microfluidics can be combined to create chemical gradients or shearing forces that cells experience in vivo. In all these cases, BioMEMS provides a powerful way to manipulate physical forces and chemical factors that interact with a cell to better understand the role of a cell's microenvironment.

Membrane Wrinkling

A pioneering development that laid the groundwork for future BioMEMS devices was Albert Harris' work in 1980. He and his colleagues observed that cells exerted traction forces by culturing cells on a deformable, silicone membrane substrate. To make the surface, a layer of silicone fluid, polydimethylsiloxane (PDMS), was spread out on a glass coverslip. Flame was then used to cross-link the top layer of the silicone to create a thin, elastic film on top of the remaining liquid silicone film, which acted as a support and lubricant for deflection. Cells placed on this film distorted the top surface and caused wrinkles that were due to the tangential forces they exerted from actin and myosin contraction. By correlating the membrane wrinkle length with a magnitude of force, the technique was later able to provide a closer measure of the traction forces, but the approach was still semiquantitative. An important finding from Harris' pioneering work was that fibroblast forces were far larger than what is needed for locomotion, in fact two or three magnitudes more. This led to the theory that these strong forces are necessary for the physical remodeling of the ECM. When fibroblasts were seeded onto collagen gels, they observed that the traction forces of the cells pulled the gel into a dense capsule of collagen fibers similar to the formation of wrinkles in the silicone films. They theorized that the traction fields created by the cells can form the aligned collagen fiber structures seen in tissues such as tendons or ligament, rather than relying on an existing matrix to induce the alignment.

A limitation of PDMS wrinkling films is that there is low resolution in measuring traction forces because there exists a mechanical coupling between a wrinkle in a substrate and the traction forces nearby. A traction force is applied at a single focal adhesion of a cell, but the surface strains in the film leads to wrinkling over a wider area. The overlap in strains from nearby traction forces makes it difficult to determine how much each one contributes to a visible wrinkle in the film. The elasticity of the silicone film also means that the traction forces of a cell can cause a strain in the film over a widespread area, which can subject neighboring cells to an external stretch or wrinkled surface topology. Quantifying wrinkles is also a vague approach because it is difficult to interpret what is and what is not a wrinkle due to debris or surface defects in the silicone film.

Refinements weremade to the silicone wrinkling technique by using surface markers that enabled a higher resolution and a better understanding of the deformations caused by traction forces. Marker beads were placed on top of the PDMS film to track displacement of the film over a high number of spatial points. To create this substrate, one micrometer diameter latex beads were deposited onto the PDMS fluid and its top surface was cross-linked in a glow discharge chamber. An improved mathematical and statistical model for the deformation of the film used the tracking of the beads, rather than wrinkles. Additionally, the approach allowed for an order of magnitude improvement in detectable forces compared to wrinkling membranes.

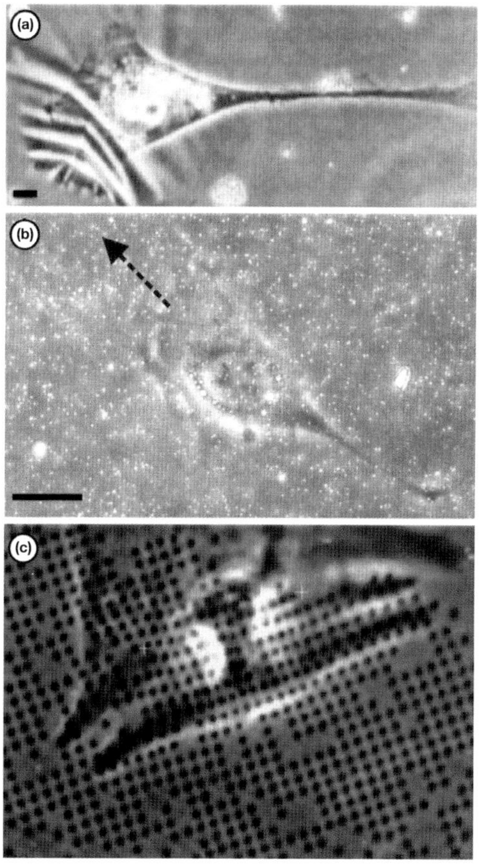

(a) Cell wrinkling of a silicone membrane floating on silicone fluid. Force is orthogonal to the long axis of each wrinkle and the stretch of the traction force correlates with the length of the wrinkles. Scale bar is 10 mm. (b) Traction force microscopy technique with cell on polyacrylamide gel containing embedded fluorescent beads, which act as fiduciary markers of strain at the surface of the substrate. Scale bar is 20 mm. (c) BioMEMS traction force microscopy where marker dots are patterned into a 2-mm grid pitch. Some dots are missing from the fabrication process, but surface strains are evident by the displacements of the markers from their original positions as evenly spaced rows and columns.

BioMEMS Devices to Apply Forces to Cells

Cells sense forces and modulate their cytoskeleton and regulatory proteins accordingly. The three main cytoskeletal filaments actin, intermediate filaments, and microtubules transmit forces within a cell, and can become linked to neighboring cells cytoskeleton through transmembrane junction proteins such as at adherence junctions. Cells therefore relay forces to cells around them, as well as experience forces acting on them from their microenvironment. The importance of these transmitted forces in tissue is important in the development of the cardiovascular, musculoskeletal, and nervous systems, all of which have been shown to be influenced by mechanical forces.

Arterial remodeling provides an interesting example of cellular mechanotransduction. Arteries are mechanically compliant, and when blood is driven through them, they expand elastically with each pressure wave from the cardiac cycle. The distensible nature of the arterial tissue acts as a stretchable substrate on which endothelial cells adhere. Strains up to 10% are normal for arterial tissue, and these strains can be recreated in vitro by culturing cells on flexible PDMS membranes that are stretched uniaxially. Cells have been observed to change their shape and realign their actin cytoskeletons in a direction perpendicular to the applied strain. However, when Rho activity is inhibited, these cells align in a direction parallel to the stretching, suggesting that external force is sufficient to induce actin remodeling.

Mechano transduction has been hypothesized to occur at a variety of cellular structures, but the cytoskeleton has been strongly implicated to have a central role since it connects to the focal adhesions and adherence junctions. When cells on a PDMS membrane were treated with a detergent to keep the cytoskeletons intact but strip away the lipid membranes and cytoplasmic proteins, it was found that cytoskeletons that were stretched had significantly higher amounts of focal adhesion proteins. This response is similar to the growth response at the focal adhesion due to traction forces. Additionally, the protein response at the focal adhesion suggests that mechanotransduction occurs from conformation changes. Force may cause unfolding of focal adhesion proteins to reveal new domains for the binding of adaptor or signaling proteins or it may also expose enzymatic domains that then activate signaling activity associated with mechanotransduction.

BioMEMS Reactors

BioMEMS microfluidic devices are frequently used to apply moving fluid to cells. To do so, a channel is formed with biocompatible materials and functionalized. Channel shape, dimensions, and flow rates are all parameters that can be adjusted to control the flow dynamics. A proven and well documented method of channel fabrication is through PDMS replication molding of channels on a silicon wafer created through surface micromachining or photolithography techniques. Complex geometries can be fabricated with this method, and valves and multiple flow conditions can be incorporated into one device. The material used to fabricate the device is flexible in many cases. For most applications, glass is the desired material due to its optical clarity, robustness, nonporous nature, and resistance to all sterilization techniques. However, machining glass and maintaining optical clarity and surface smoothness is a difficult task. Other common biocompatible replacement materials are acrylic, polystyrene, and polycarbonate. These offer similar optical transparency, can still be sterilized, and are easy to reconfigure in a laboratory setting. Plasma treatment of glass or a polymer creates a highly adhesive surface that forms a watertight seal with PDMS and is employed in many bioreactor designs. For live-cell observation, a glass coverslip makes an ideal bottom surface of a bioreactor due to its thinness, a requirement for high magnification microscope objectives.

Cells are introduced to microfluidic systems by suspending them in media and injecting media through the micro channel. The cells settle and adhere to the surface and washing steps remove unattached cells. Since the vascular wall is comprised of more than endothelial cells, researchers have examined the effects of co culture on alignment and adhesion. First, a layer of smooth muscle cells is seeded onto a substrate and grown to quiescent confluence. Next, endothelial cells can be directly seeded on top of the first layer and allowed to proliferate to confluence. The result is a bilayer of cells that closely resembles the endothelium in structure. The two cell types are able to form trans membrane gap junctions for chemical signaling, as well as respond to mechanical stresses through adherence junctions.

Bioreactor Pumping

To create fluid flow, different pumping schemes are used to achieve the desired flow in the microchannels. In the most basic flow device, pressure produced either through a raised fluid reservoir or gas source can drive fluid through a microfluidic system while a pump cycles the fluid back into the reservoir source. However, the lack of feedback-control means that researchers require more precise means. Syringe pumps provide excellent control if low flow rates are needed, but with limited volumes. Configuring multiple syringes and valves together can create a unidirectional flow that can recirculate to maintain the supply volume. Large flow rates use a peristaltic pump, also known as a roller pump, to have precise control of the flow. Peristaltic pumps produce a nonsteady pressure to the flow, which can be desirable if researchers want pulsatile flow. In most cases, a steady flow rate is desired, so air dampeners are placed in the fluid path to remove the effect of the pulses. on other types of fluid flow devices.

Bioreactor Configuration

In BioMEMS microfluidic applications, ducts are the most common channel shape because shear is well defined and can be controlled precisely by dimensional considerations. Certain assumptions about the length to-width aspect ratio mean that the wall shear stresses in the microchannels can be approximated to shear stress in parallel-plate flow. Shear forces at the wall can be calculated using: $\tau = 6\,\mu Q / WH^2$, where τ is shear stress, μ is viscosity, Q is fluid volumetric flow rate, W is the width, and H is the height of the channel. With steady pumping schemes, a constant level of shear stress gradient is produced at the wall surface. Adding a weir or obstruction in the channel initiates a disturbance in the flow that causes recirculation or vorticity at a region downstream. Between the weir and the vorticity reattachment point, a range of shear stresses and shear stress gradient exists.

References

- Galajda P; J.E. Keymer; P Chaikin; R. Austin (December 2007). "A Wall of Funnels Concentrates Swimming Bacteria". Journal of Bacteriology. 189 (23): 8704–8707. doi:10.1128/JB.01033-07. PMC 2168927 . PMID 17890308.
- Global-mems-gyroscopes-market: gosreports.com, Retrieved 13 May 2018
- Herold, KE; Rasooly, A (editor) (2009). Lab-on-a-Chip Technology: Biomolecular Separation and Analysis. Caister Academic Press. ISBN 978-1-904455-47-9.
- Kaiser, R. I.; Stockton, A. M.; Kim, Y. S.; Jensen, E. C.; Mathies, R. A. (2013). "On the Formation of Dipeptides in Interstellar Model Ices". The Astrophysical Journal. 765 (2):
- Digital-micromirror-device: definedterm.com, Retrieved 30 March 2018
- Angelani L.; R. Di Leonardo; G. Ruocco (2009). "Self-Starting Micromotors in a Bacterial Bath". Phys. Rev. Lett. 102: 048104. arXiv:0812.2375 . Bibcode:2009PhRvL.102d8104A. doi:10.1103/PhysRevLett.102.048104.
- Ahmed, T.; Shimizu, T.S.; Stocker, R. (2010). "Microfluidics for bacterial chemotaxis". Integrative Biology. 2: 604–629. doi:10.1039/C0IB00049C.
- What-is-Microfluidics, life-sciences: news-medical.net, Retrieved 10 July 2018
- Berthier, Jean; Brakke, Kenneth A.; Berthier, Erwin (2016). Open Microfluidics. John Wiley & Sons, Inc. pp. 229–256. doi:10.1002/9781118720936.ch7/summary. ISBN 9781118720936.

Permissions

All chapters in this book are published with permission under the Creative Commons Attribution Share Alike License or equivalent. Every chapter published in this book has been scrutinized by our experts. Their significance has been extensively debated. The topics covered herein carry significant information for a comprehensive understanding. They may even be implemented as practical applications or may be referred to as a beginning point for further studies.

We would like to thank the editorial team for lending their expertise to make the book truly unique. They have played a crucial role in the development of this book. Without their invaluable contributions this book wouldn't have been possible. They have made vital efforts to compile up to date information on the varied aspects of this subject to make this book a valuable addition to the collection of many professionals and students.

This book was conceptualized with the vision of imparting up-to-date and integrated information in this field. To ensure the same, a matchless editorial board was set up. Every individual on the board went through rigorous rounds of assessment to prove their worth. After which they invested a large part of their time researching and compiling the most relevant data for our readers.

The editorial board has been involved in producing this book since its inception. They have spent rigorous hours researching and exploring the diverse topics which have resulted in the successful publishing of this book. They have passed on their knowledge of decades through this book. To expedite this challenging task, the publisher supported the team at every step. A small team of assistant editors was also appointed to further simplify the editing procedure and attain best results for the readers.

Apart from the editorial board, the designing team has also invested a significant amount of their time in understanding the subject and creating the most relevant covers. They scrutinized every image to scout for the most suitable representation of the subject and create an appropriate cover for the book.

The publishing team has been an ardent support to the editorial, designing and production team. Their endless efforts to recruit the best for this project, has resulted in the accomplishment of this book. They are a veteran in the field of academics and their pool of knowledge is as vast as their experience in printing. Their expertise and guidance has proved useful at every step. Their uncompromising quality standards have made this book an exceptional effort. Their encouragement from time to time has been an inspiration for everyone.

The publisher and the editorial board hope that this book will prove to be a valuable piece of knowledge for students, practitioners and scholars across the globe.

Index